Contents

TABLES

BOXES

FIGURES

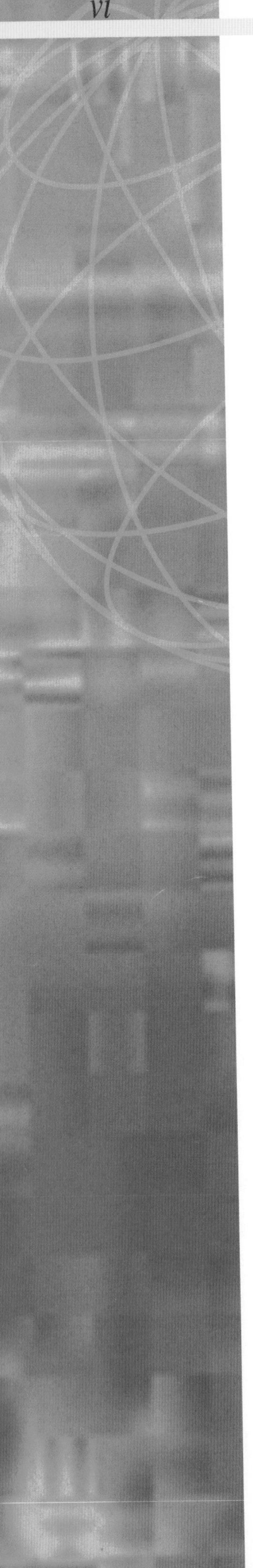

Foreword

Every era has its challenges. And each challenge demands specific responses.

In the 1960s, famine threatened South Asia. The Green Revolution was the right answer to the looming food crisis that the world faced half a century ago.

Fortunately, today we are not facing the prospect of large-scale famine – but we are at a crossroads.

About 842 million people remain chronically hungry because they cannot afford to eat adequately, despite the fact that the world is no longer short of food. In a disconcerting paradox, more than 70 percent of the world's food-insecure people live in rural areas in developing countries. Many of them are low-paid farm labourers or subsistence producers who may have difficulty in meeting their families' food needs.

As we look towards 2050, we have the additional challenge of feeding a population that is eating more – and sometimes better and healthier diets – and that is expected to surpass the 9 billion mark. At the same time, farmers, and humanity as a whole, are already facing the new challenges posed by climate change. The widespread degradation of land and water resources, along with other negative environmental impacts, is showing us the limits of highly intensive farming systems.

Hence, the quest is now to find farming systems that are truly sustainable and inclusive and that support increased access for the poor so that we can meet the world's future food needs. Nothing comes closer to the sustainable food production paradigm than family farming.

It is therefore fitting that the United Nations has declared 2014 the International Year of Family Farming. This provides an occasion to highlight the role that family farmers – a sector that includes small and medium-scale farmers, indigenous peoples, traditional communities, fishers, pastoralists, forest dwellers, food gatherers and many others – play in food security and sustainable development.

To celebrate the International Year of Family Farming, *The State of Food and Agriculture 2014: Innovation in family farming* (SOFA 2014) offers a groundbreaking study of family farming. The report contains the first comprehensive estimate of the number of family farms in the world – at least 500 million. This means that families run about nine out of ten farms. Additional analysis shows that family farms occupy a large share of the world's agricultural land and produce about 80 percent of the world's food.

However, while family farmers are key to food security worldwide, they have also been considered by many as an obstacle to development and have been deprived of government support. That is the mindset we need to change. Family farmers are not part of the problem: on the contrary, they are vital to the solution of the hunger problem.

But there is a limit to what family farmers can achieve on their own, and the role of the public sector is to put in place the policies and create the enabling environment that will enable them to flourish. This must be a government-led effort, but is one that calls for the participation of others as well: international organizations, regional agencies, civil society organizations, the private sector and research institutions.

The sheer diversity of family farms and the complexity of their livelihoods mean that one-size-fits-all recommendations are not appropriate. In supporting family farms, each country and each region needs to find the solutions that best respond to family farmers' specific needs and the local context and that build on family farmers' inherent capacities and strengths.

However, what family farmers need is broadly similar throughout the world: improved access to technologies that bolster sustainable increases in productivity without unduly raising risks; inputs that respond to their particular needs and respect their cultures and traditions; special attention to women and young farmers; strengthened producers' organizations and cooperatives; improved access to land and water, credit and markets; improved participation in value chains, including an assurance of fair prices;

strengthened links between family farming and local markets to increase local food security; and equitable access to essential services including education, health, clean water and sanitation.

At the same time, support to family farmers must underpin their role in promoting development in rural communities. Beyond increasing local food availability, family farmers play a vital role in creating jobs, generating income and stimulating and diversifying local economies.

There are many ways through which we can nurture this potential. These include linking family farming production to institutional markets destined, for instance, to supply school meals – a combination that guarantees markets and income to family farmers and nutritious meals for children. Family farmers are also well placed to recover traditional crops that have great value for local food security but that have been left aside because of the *commodification* of our diets.

There is a wealth of successful experiences from around the world that can serve as examples to other countries in bringing about the changes needed to fulfil the potential of their family farmers. SOFA 2014 outlines options for responding to the needs of and opportunities for family farmers in different contexts.

These options all have a common feature: innovation. Family farmers need to innovate in the systems they use; governments need to innovate in the specific policies they implement to support family farming; producers' organizations need to innovate to respond better to the needs of family farmers; and research and extension institutions need to innovate by shifting from a research-driven process predominantly based on technology transfer to an approach that enables and rewards innovation by family farmers themselves. Additionally, in all its forms, innovation needs to be inclusive, involving family farmers in the generation, sharing and use of knowledge so that they have ownership of the process, taking on board both the benefits and the risks, and making sure that it truly responds to local contexts.

We need a way forward that is as innovative as the Green Revolution was but that responds to today's needs and looks to the future: we cannot use the same tool to respond to a different challenge.

The 2014 International Year of Family Farming reminds us of the need to act to revitalize this critical sector. By choosing to celebrate family farmers, we recognize that they are natural leaders in the response to the three big challenges facing the farming world today: improving food security and nutrition while preserving crucial natural resources and limiting the extent of climate change.

If we give family farmers the attention and support they need and deserve, together we can rise to these challenges.

José Graziano da Silva
FAO Director-General

Acknowledgements

The State of Food and Agriculture 2014 was prepared by members of FAO's Agricultural Development Economics Division (ESA) and the Research and Extension Unit (DDNR) under the overall leadership of Kostas Stamoulis, Director of ESA, Andrea Sonnino, Chief of DDNR and Terri Raney, Senior Economist and Chief Editor (ESA). Additional guidance was provided by Jomo Kwame Sundaram, Assistant Director-General of the Economic and Social Development Department.

The research and writing team was led by Jakob Skoet (ESA) and David Kahan (DDNR) and included: Brian Carisma, Sarah Lowder, Sara McPhee Knowles and Terri Raney (ESA); John Ruane and Julien de Meyer (DDNR). Several other FAO colleagues provided inputs to the report: Aslihan Arslan, Solomon Asfaw, Panagiotis Karfakis, Leslie Lipper, Giulia Ponzini, George Rapsomanikis and Saumya Singh (ESA); Magdalena Blum, Delgermaa Chuluunbaatar, Steven LeGrand, Karin Nichterlein, Ana Pizarro and Laura Vian (DDNR); May Hani, Social Protection Division; Adriana Neciu and Jairo Castano, Statistics Division; Manuela Allara and Benjamin Graeub, Plant Production and Protection Division; Nora Ourabah Haddad and Denis Herbel, Office for Partnerships, Advocacy and Capacity Development; John Preissing, FAO–Peru; and Stephen Rudgard, FAO–Laos. Many other FAO colleagues from various technical divisions and regional offices provided expert reviews and advice on multiple drafts of the report, and their contributions are gratefully acknowledged. External background papers and inputs were prepared by: Ian Christoplos, Glemminge Development Research; Keith Fuglie, Economic Research Service, US Department of Agriculture; Silvia L. Saravia Matus, independent consultant; Philip G. Pardey, University of Minnesota; and Helena Posthumus, Royal Tropical Institute of the Netherlands (KIT).

The report benefited from external reviews and advice from many international experts: Nienke Beintema, José Falck-Zepeda and Keith Wiebe, International Food Policy Research Institute (IFPRI); Mark Holderness and Thomas Price, Global Forum on Agricultural Research (GFAR); Kristin Davis, Global Forum for Rural Advisory Services (GFRAS); Helen Hambly Odame, University of Guelph; Laurens Klerkx; University of Wageningen; Donald Larson, World Bank; Moses Makooma Tenywa, Makerere University; Gigi Manicad, Oxfam Novib; Hannington Odame, Centre for African Bio-Entrepreneurship (CABE); Bernard Triomphe, Agricultural Research Centre for Development (CIRAD); and Xiangping Jia, Center for Chinese Agricultural Policy, Chinese Academy of Sciences.

Initial guidance for the study from participants of the FAO Expert Consultation on agricultural innovation systems and family farming (March 2012) are gratefully acknowledged, as well from participants of the subsequent e-mail conference on the same theme (June–July 2012), which was managed by John Ruane (DDNR).

Mariana Wongtschowski, Royal Tropical Institute of the Netherlands (KIT) facilitated the technical review workshop, which discussed and reviewed the first comprehensive draft of the report. Michelle Kendrick, Economic and Social Development Department, was responsible for publishing and project management. Paola Landolfi assisted the production cycle. Paola Di Santo, Liliana Maldonado and Cecilia Agyeman-Anane provided administrative support and Marco Mariani arranged for IT support throughout the process. Editing was carried out by Jane Shaw. Translation and printing services were delivered by the FAO Meeting Programming and Documentation Service. Graphic design and layout services were supplied by Flora Dicarlo.

Abbreviations and acronyms

CGIAR	Consultative Group on International Agricultural Research
FFS	Farmer Field School
G20	Group of Twenty Finance Ministers and Central Bank Governors
GDP	gross domestic product
ICT	information and communication technology
IFAD	International Fund for Agricultural Development
IFPRI	International Food Policy Research Institute
MAFAP	Monitoring African Food and Agricultural Policies
NGO	non-governmental organization
OECD	Organisation for Economic Co-operation and Development
R&D	research and development
TAP	Tropical Agricultural Platform

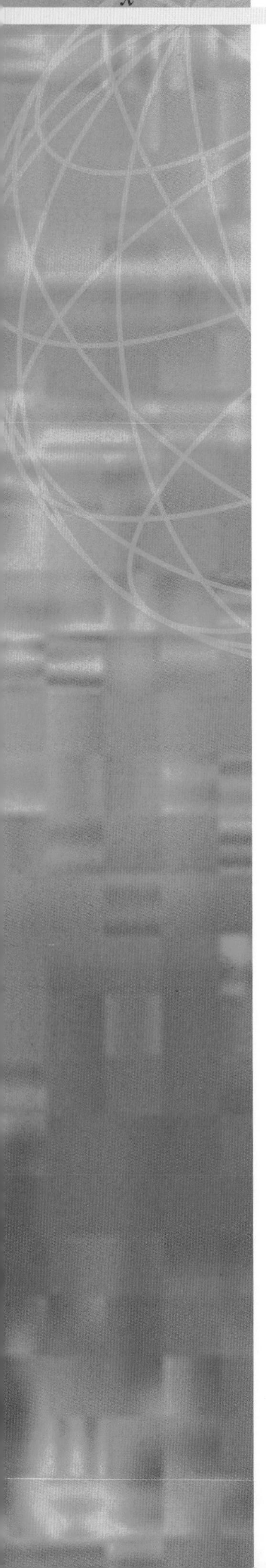

Executive summary

Innovation in family farming

More than 500 million family farms manage the majority of the world's agricultural land and produce most of the world's food. We need family farms to ensure global food security, to care for and protect the natural environment and to end poverty, undernourishment and malnutrition. These goals can be thoroughly achieved if public policies support family farms to become more productive and sustainable; in other words policies must support family farms to innovate within a system that recognizes their diversity and the complexity of the challenges faced.

The State of Food and Agriculture 2014: Innovation in family farming analyses family farms and the role of innovation in ensuring global food security, poverty reduction and environmental sustainability. It argues that family farms must be supported to innovate in ways that promote sustainable intensification of production and improvements in rural livelihoods. Innovation is a process through which farmers improve their production and farm management practices. This may involve planting new crop varieties, combining traditional practices with new scientific knowledge, applying new integrated production and post-harvest practices or engaging with markets in new, more rewarding ways. But innovation requires more than action by farmers alone. The public sector – working with the private sector, civil society and farmers and their organizations – must create an innovation system that links these various actors, fosters the capacity of farmers and provides incentives for them to innovate.

Family farms are very diverse in terms of size, access to markets and household characteristics, so they have different needs from an innovation system. Their livelihoods are often complex, combining multiple natural-resource-based activities, such as raising crops and animals, fishing, and collecting forest products, as well as off-farm activities, including agricultural and non-agricultural enterprises and employment. Family farms depend on family members for management decisions and most of their workforce, so innovation involves gender and intergenerational considerations. Policies will be more effective if they are tailored to the specific circumstances of different types of farming households within their institutional and agro-ecological settings. Inclusive research systems, advisory services, producer organizations and cooperatives, as well as market institutions are essential.

The challenges of designing an innovation system for the twenty-first century are more complex than those faced at the time of the Green Revolution. The institutional framework is different due to a declining role of the public sector in agricultural innovation and the entry of new actors, such as private research companies and advisory services, as well as civil society organizations. At the same time, farmers are having to address globalization, increasingly complex value chains, pressures on natural resources, and climate change.

Family farms: size and distribution*

There are more than 570 million farms in the world. Although the notion of family farming is imprecise, most definitions refer to the type of management or ownership and the labour supply on the farm. More than 90 percent of farms are run by an individual or a family and rely primarily on family labour. According to these criteria, family farms are by far the most prevalent form of

* Assessing the number of farms and family farms as well as land distribution throughout the world is difficult because of the absence of systematic and comparable data for all countries. Estimates presented here are based on agricultural censuses for different time periods and different countries, and are intended to provide indications of orders of magnitude rather than exact numbers.

agriculture in the world. Estimates suggest that they occupy around 70–80 percent of farm land and produce more than 80 percent of the world's food in value terms.

The vast majority of the world's farms are small or very small, and in many lower-income countries farm sizes are becoming even smaller. Worldwide, farms of less than 1 hectare account for 72 percent of all farms but control only 8 percent of all agricultural land. Slightly larger farms between 1 and 2 hectares account for 12 percent of all farms and control 4 percent of the land, while farms in the range of 2 to 5 hectares account for 10 percent of all farms and control 7 percent of the land. In contrast, only 1 percent of all farms in the world are larger than 50 hectares, but these few farms control 65 percent of the world's agricultural land. Many of these large, and sometimes very large, farms are family-owned and operated.

The highly skewed pattern of farm sizes at the global level largely reflects the dominance of very large farms in high-income and upper-middle-income countries and in countries where extensive livestock grazing is a dominant part of the agricultural system. Land is somewhat more evenly distributed in the low- and lower-middle-income countries, where more than 95 percent of all farms are smaller than 5 hectares. These farms occupy almost three-quarters of all farm land in the low-income countries and almost two-thirds in the lower-middle-income group. In contrast, farms larger than 50 hectares control only 2 percent and 11 percent, respectively, of the land in these income groups.

Exactly what can be considered a small farm – below 0.5 or 1 hectare, or some other size – will depend on agro-ecological and socio-economic conditions, and their economic viability will depend on market opportunities and policy choices. Below a certain level, a farm may be too small to constitute the main means of support for a family. In this case, agriculture may make an important contribution to a family's livelihood and food security, but other sources of income through off-farm employment, transfers or remittances are necessary to ensure the family lives a decent life. On the other hand, many small or medium-sized family farms in the low- and middle-income countries could make a greater contribution to global food security and rural poverty alleviation, depending on their productive potential, access to markets and capacity to innovate. Through a supportive agricultural innovation system these farms could help transform world agriculture.

Family farms, food security and poverty

In most countries, small and medium-sized farms tend to have higher agricultural crop yields per hectare than larger farms because they manage resources and use labour more intensively. This means that the share of small and medium-sized farms in national food production is likely to be even larger than the share of land they manage.

A large proportion of family farmers with small landholdings also depend on other natural resources, especially forests, pastureland and fisheries. The intensive resource use on these farms may threaten sustainability of production. These small and medium-sized farms are central to global natural resource management and environmental sustainability as well as to food security.

While smaller farms tend to achieve higher yields per hectare than larger farms, they produce less per worker. Labour productivity – or output per worker – is also much lower in low-income countries than in high-income countries. Increased labour productivity is a precondition for sustained income growth, so enabling farming families in low- and middle-income countries to raise their labour productivity is essential if we are to boost farm incomes and make inroads into reducing rural poverty.

Although smaller farms tend to have higher yields than larger farms within the same country, cross-country comparisons show that yields per hectare are much lower in poorer countries, where smaller farms are more prevalent, than in richer countries. This seeming paradox simply reflects the fact that yields in low-income countries are far lower, on average, than in richer countries and far lower than they could be if existing

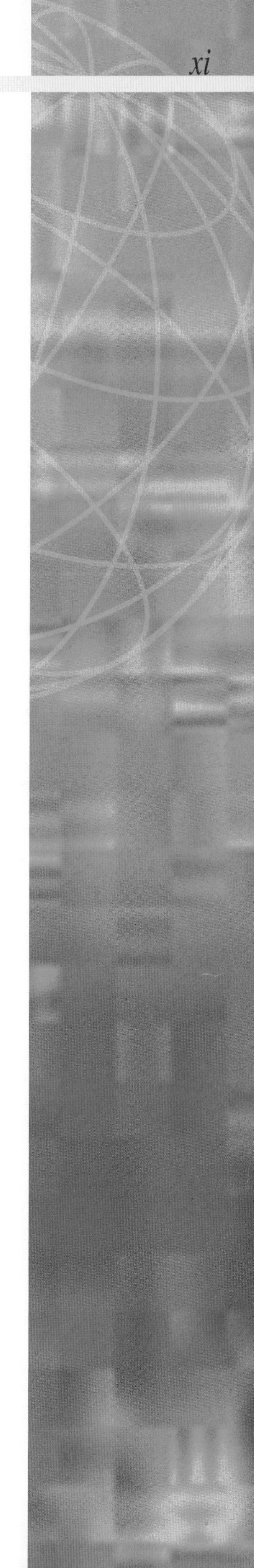

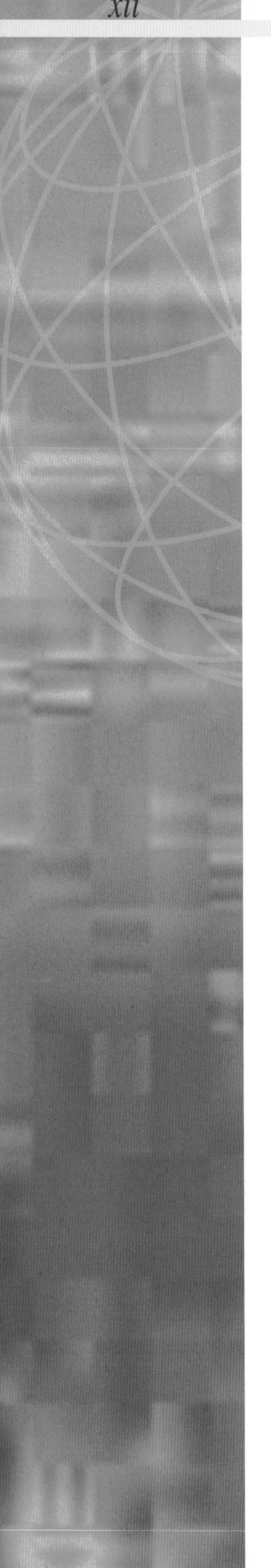

technologies and management practices were appropriately adapted and more widely adopted in low-income countries. Innovation aimed at increasing yields in developing countries could have significant impacts in terms of expanding agricultural production, increasing farm incomes and lowering food prices, thereby reducing poverty and enhancing food security by making food more affordable and accessible to both rural and urban populations.

The potential to improve labour productivity and yields can only be realized if family farmers are able to innovate. There are two main, but interrelated, pathways through which farmers' productivity may be increased: the development, adaptation and application of new technologies and farm management practices; and the wider application of existing technologies and practices. The first expands the potential for more productive use of existing resources by pushing out the production possibility frontier. The second allows farmers to achieve more of this potential.

Innovation systems for family farming

Innovation happens when individuals and groups adopt new ideas, technologies or processes that, when successful, spread through communities and societies. The process is complex, involving many actors, and it cannot function in a vacuum. It is furthered by the presence of an effective *innovation system*. Among other things, an agricultural innovation system includes the general enabling economic and institutional environment required by all farmers. Other key components are research and advisory services and effective agricultural producers' organizations. Innovation often builds on and adjusts local knowledge and traditional systems in combination with new sources of knowledge from formal research systems.

One fundamental driver for all innovators – including family farmers – is access to markets that reward their enterprise. Farmers with access to markets, including local markets, for their produce – whether it be food staples or cash crops – have a strong incentive to innovate. Technologies help farmers to enter the market by allowing them to produce marketable surpluses. Innovation and markets depend on, and reinforce, each other. However, investments in physical and institutional market infrastructure are essential to allow farmers to access markets both for their produce and for inputs. Efficient producers' organizations and cooperatives can also play a key role in helping farmers link to input and output markets.

Because family farms are so diverse in terms of size, access to markets and other characteristics, general policy prescriptions are unlikely to meet the needs of all of them. Public support for innovation should take into consideration the specific structure of family farming in each country and setting, as well as the policy objectives for the sector.

Some family farmers manage large commercial enterprises and require little from the public sector beyond agricultural research to ensure long-term production potential and the enabling environment and infrastructure that all farmers need to be productive, although they may require regulation, support and incentives to become more sustainable. Other, very small, family farms engage in markets primarily as net food buyers. They produce food as an essential part of their survival strategy, but they often face unfavourable policy environments and have inadequate means to make farming a commercially viable enterprise. Many such farmers supplement both income and nutrition from other parts of the landscape, through forests, pastures and fisheries and from off-farm employment. For these farmers, diversification and risk spreading through these and other livelihood strategies will be necessary. While agriculture and agricultural innovation can improve livelihoods, they are unlikely to be the primary means of lifting this group of farmers out of poverty. Helping such farmers escape poverty will require broad-based efforts, including overall rural development policies and effective social protection. In between these two extremes are the millions of small and medium-sized family farms that have the potential to become economically viable and environmentally sustainable enterprises. Many of these farms are not well integrated into effective innovation systems and lack the capacity or incentives to innovate.

Public efforts to promote innovation in agriculture for family farms must focus on providing inclusive research, advisory services, market institutions and infrastructure that the private sector is typically unable to provide. For example, applied agricultural research for crops, livestock species and management practices of importance to smallholders are public goods and should be a priority. A supportive environment for producer organizations and other community-based organizations can also help promote innovation among family farms.

Promoting sustainable productivity on family farms

Demand for food is growing while land and water resources are becoming ever more scarce and degraded. Climate change will make these challenges yet more difficult. Over the coming decades, farmers need to produce significantly larger amounts of food, mostly on land already in production. The large gaps between actual and potential yields for major crops show that there is significant scope for increased production through productivity growth on family farms. This can be achieved by developing new technologies and practices or through overcoming barriers and constraints to the adaptation and adoption of existing technologies and practices. Overcoming poverty in low- and middle-income countries also means boosting labour productivity through innovation on family farms as well as providing farming families with other opportunities for employment.

It is not enough to produce more. If societies are to flourish in the long term, they must produce sustainably. The past paradigm of input-intensive production cannot meet the challenge. Productivity growth must be achieved through sustainable intensification. That means, *inter alia*, conserving, protecting and enhancing natural resources and ecosystems, improving the livelihoods and well-being of people and social groups and bolstering their resilience – especially to climate change and volatile markets.

The world must rely on family farms to grow the food it needs and to do so sustainably. For this to happen, family farmers must have the knowledge and economic and policy incentives they need to provide key environmental services, including watershed protection, biodiversity conservation and carbon sequestration.

Overcoming barriers to sustainable farming

Smaller family farms tend to rely on tried and trusted methods because one wrong decision can jeopardize an entire growing season; but they readily adopt new technologies and practices that they perceive to be beneficial in their specific circumstances. Nevertheless, several obstacles often stand in the way of farmers adopting innovative practices that combine productivity increases with preservation and improvement of natural resources. Key impediments include the absence of physical and marketing infrastructure, financial and risk management instruments, and secure property rights.

Farmers often face high initial costs and long pay-off periods when making improvements. This can prove to be a prohibitive disincentive, especially in the absence of secure land rights and of access to financing and credit. Farmers are also unlikely to undertake costly activities and practices that generate public goods (such as environmental conservation) without compensation or local collective action. Furthermore, improved farm practices and technologies often only work well in the agro-ecological and social contexts for which they were designed, and if solutions are not adapted to local conditions, this can be a serious impediment to adoption.

Local institutions, such as producers' organizations, cooperatives and other community-based organizations, have a key role to play in overcoming some of these

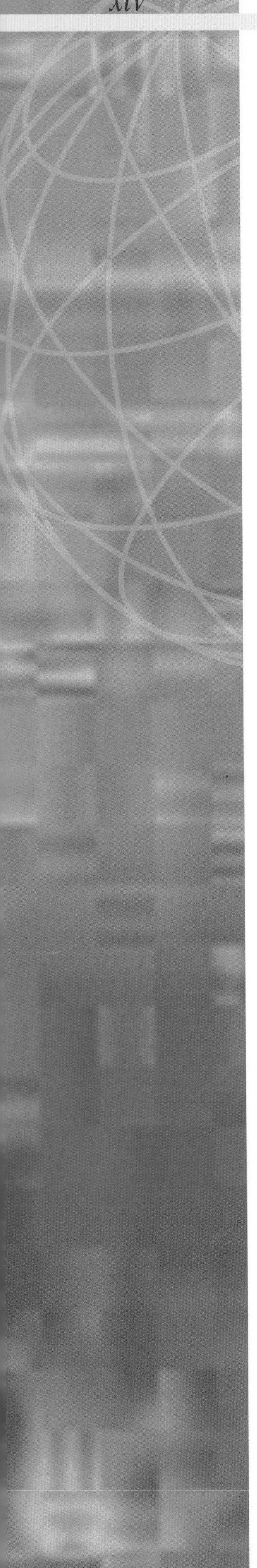

barriers. The effective functioning of local institutions and their coordination with the public and private sectors and with farmers themselves, both men and women, can determine whether or not small family farms can introduce innovative, sustainable improvements suited to their needs and local conditions.

Agricultural research and development – focusing on family farms

Investing in agricultural research and development (R&D) is important for boosting agricultural productivity, preserving the environment and eradicating poverty and hunger. A large body of evidence confirms that there are high returns to public investments in agricultural R&D. In many countries such investment is currently insufficient. Private-sector research is increasingly important, especially in high-income countries, but it cannot replace public research. Much agricultural research can be considered a public good, where the benefits of the knowledge generated cannot be appropriated by a private company and is therefore unlikely to attract the private sector. Returns to agricultural R&D often take a long time to materialize and, in addition, research is cumulative, with results building up over time. In this context, a continuous long-term public commitment to agricultural research is fundamental. Innovative forms of more short-term financing can help, but stable institutional funding is needed to maintain a core long-term research capacity.

All countries need a certain level of domestic research capacity because technologies and practices can rarely be imported without some adaptation to local agro-ecological conditions. However, countries need to consider carefully what research strategy is best suited to their specific needs and capacities. Some countries, particularly those with too few funds to run strong national research programmes, may need to focus on adapting the results of international research to conditions at home. Others, with bigger research budgets, may also want to devote resources to more basic research. The establishment of international partnerships and a careful division of labour between international research with broader applications and national research geared to domestic needs is a priority. There is also scope for South–South cooperation between large countries with major public research programmes and countries with less national research capacity facing similar agro-ecological conditions.

Research that meets the needs of family farms in their specific agro-ecological and social conditions is essential. Combining farmer-led innovation and traditional knowledge with formal research can contribute to sustainable productivity. Involving family farmers in defining research agendas and engaging them in participatory research efforts can improve the relevance of research for them. This may include working closely with producers' organizations and creating incentives for researchers and research organizations to interact with family farms and their different members, including women and youth, and to undertake research tailored to their specific circumstances and needs.

Promoting inclusive rural advisory services

While investments in agricultural R&D are needed in order to expand the potential for sustainable production, sharing knowledge about technologies and innovative practices among family farmers is perhaps even more important for closing existing gaps in agricultural productivity and sustainability between developing and developed countries. Agricultural extension and advisory services are critical for this challenge, but far too many farmers, and especially women, do not have regular access to such services. Modern extension features many different kinds of advisory services as well as service providers from the public, private and non-profit sectors. While there is no standard model for delivery of extension services, governments, private businesses, universities, NGOs, and producer organizations can play the role of service providers for different purposes and for different approaches. Strengthening the various types of service providers is an important component of promoting innovation.

Governments still have a strong role to play in the provision of agricultural advisory services. Like research, agricultural advisory services generate benefits for society that are greater than the value captured by individual farmers and commercial advisory service providers. These benefits – increased productivity, improved sustainability, lower food prices, poverty reduction, etc. – constitute public goods and call for the involvement of the public sector in the provision of agricultural advisory services. In particular, the public sector has a clear role in providing services to small family farms, especially in remote areas, who are unlikely to be reached by commercial service providers and who may have a strong need for neutral advice and information on suitable farming practices. Other areas include the provision of advisory services relating to more sustainable agricultural practices, or for climate change adaptation or mitigation through reduced greenhouse gas emissions or increased carbon sequestration. The public sector is also responsible for ensuring that the advisory services provided by the private sector and civil society are technically sound and socially and economically appropriate.

For rural advisory services to be relevant and have the necessary impact, the needs of different types of family farms as well as different household members in farming families need to be addressed. Engaging women and youth effectively and ensuring that they have access to advisory services that take into account their needs and constraints are central to ensuring effectiveness. Participatory approaches, e.g. farmer field schools in which farmers learn from other farmers, peer-learning mechanisms and knowledge-sharing activities, provide effective means for achieving these aims. More information and evidence is needed on experiences with different extension models and their effectiveness. Efforts to gather and share such information should be promoted at the national and international levels.

Developing capacity for innovation in family farming

Innovation presupposes a *capacity to innovate* at the individual, collective, national and international levels. The skills and capacities of individuals involved in all aspects of the agricultural innovation system – farmers, extension service providers, researchers, etc. – must be upgraded through education and training at all levels. Special attention needs to be given to women and girls based on their needs and roles in agriculture and rural livelihood strategies. A further focus must also be on youth in general, who tend to have a greater inclination to innovate than elder farmers and represent the future of agriculture. If youth perceive agriculture as a potential profession with scope for innovation, this can have major positive implications for the prospects for the sector.

Collective innovation capacity depends on effective networks and partnerships among the individuals and groups within the system. Producers' organizations and cooperatives are of particular importance. Strong, effective and inclusive organizations can facilitate the access of family farms to markets for inputs and outputs, to technologies and to financial services such as credit. They can serve as a vehicle for closer cooperation with national research institutes; provide extension and advisory services to their members; act as intermediaries between individual family farms and different information providers; and help small farmers gain a voice in policy-making to counter the often prevailing influence of larger, more powerful interests. Furthermore, family farmers who depend on other resources, such as forests, pastures and fisheries can benefit by linking with producer organizations within these sectors. Linking producer organizations across these sectors can further strengthen the case for clear tenure rights and better coordination between policies and service providers.

At national and international levels, the right environment and incentives for innovation are created by good governance and sound economic policies, secure property rights, market and other infrastructure, and a conducive regulatory

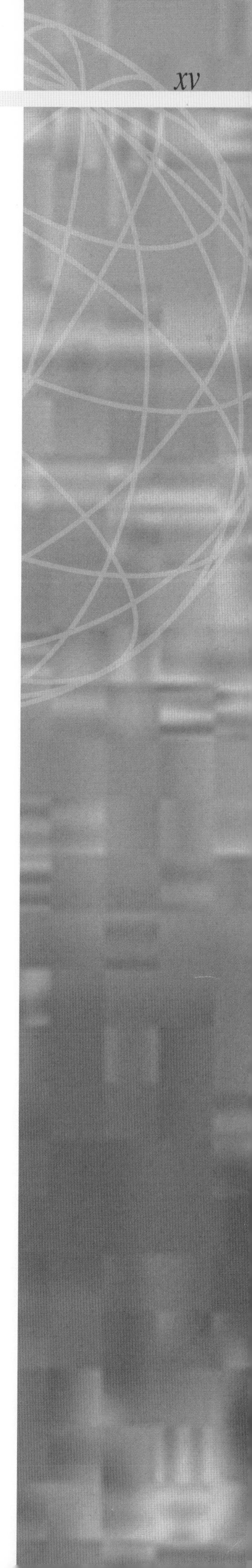

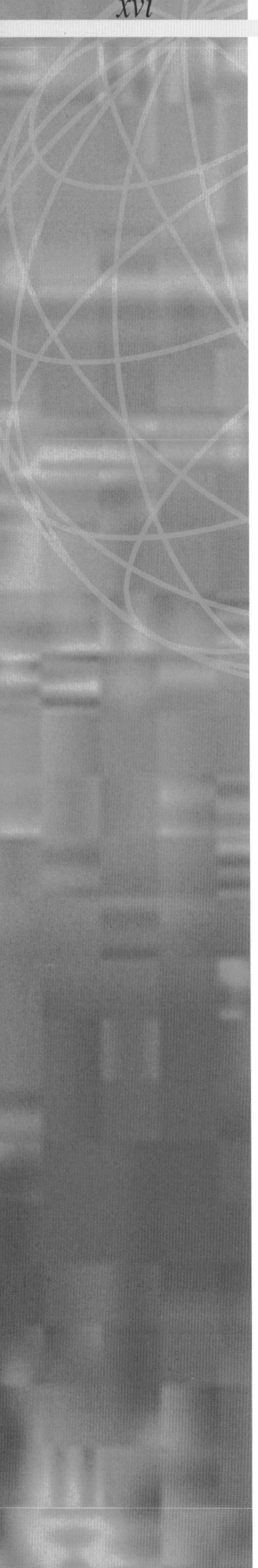

framework. Governments must support the development of effective and representative producers' organizations and ensure that they participate in policy-making processes.

Key messages of the report

- **Family farms are part of the solution for achieving food security and sustainable rural development; the world's food security and environmental sustainability depend on the more than 500 million family farms that form the backbone of agriculture in most countries.** Family farms represent more than nine out of ten farms in the world and can serve as a catalyst for sustained rural development. They are the stewards of the world's agricultural resources and the source of more than 80 percent of the world's food supply, but many of them are poor and food-insecure themselves. Innovation in family farming is urgently needed to lift farmers out of poverty and help the world achieve food security and sustainable agriculture.
- **Family farms are an extremely diverse group, and innovation systems must take this diversity into account.** Innovation strategies for all family farms must consider their agro-ecological and socio-economic conditions and government policy objectives for the sector. Public efforts to promote agricultural innovation for small and medium-sized family farms should ensure that agricultural research, advisory services, market institutions and infrastructure are inclusive. Applied agricultural research for crops, livestock species and management practices of importance to these farms are public goods and should be a priority. A supportive environment for producers' and other community-based organizations can help promote innovation, through which small and medium-sized family farms could transform world agriculture.
- **The challenges facing agriculture and the institutional environment for agricultural innovation are far more complex than ever before; the world must create an innovation system that embraces this complexity.** Agricultural innovation strategies must now focus not just on increasing yields but also on a more complex set of objectives, including preserving natural resources and raising rural incomes. They must also take into account today's complex policy and institutional environment for agriculture and the more pluralistic set of actors engaged in decision-making. An *innovation system* that facilitates and coordinates the activities of all stakeholders is essential.
- **Public investment in agricultural R&D and extension and advisory services should be increased and refocused to emphasize sustainable intensification and closing yield and labour productivity gaps.** Agricultural research and advisory services generate public goods – productivity, improved sustainability, lower food prices, poverty reduction, etc. – calling for strong government involvement. R&D should focus on sustainable intensification, continuing to expand the production frontier but in sustainable ways, working at the system level and incorporating traditional knowledge. Extension and advisory services should focus on closing yield gaps and raising the labour productivity of small and medium-sized farmers. Partnering with producers' organizations can help ensure that R&D and extension services are inclusive and responsive to farmers' needs.
- **All family farmers need an enabling environment for innovation, including good governance, stable macroeconomic conditions, transparent legal and regulatory regimes, secure property rights, risk management tools and market infrastructure.** Improved access to local or wider markets for inputs and outputs, including through government procurement from family farmers, can provide strong incentives for innovation, but farmers in remote areas and marginalized groups often face severe barriers. In addition, sustainable agricultural practices often

have high start-up costs and long pay-off periods and farmers may need appropriate incentives to provide important environmental services. Effective local institutions, including farmers' organizations, combined with social protection programmes, can help overcome these barriers.

- **Capacity to innovate in family farming must be promoted at multiple levels.** Individual innovation capacity must be developed through investment in education and training. Incentives are needed for the creation of networks and linkages that enable different actors in the innovation system – farmers, researchers, advisory service providers, value chain participants, etc. – to share information and work towards common objectives.
- **Effective and inclusive producers' organizations can support innovation by their members.** Producers' organizations can assist their members in accessing markets and linking with other actors in the innovation system. They can also help family farms have a voice in policy-making.

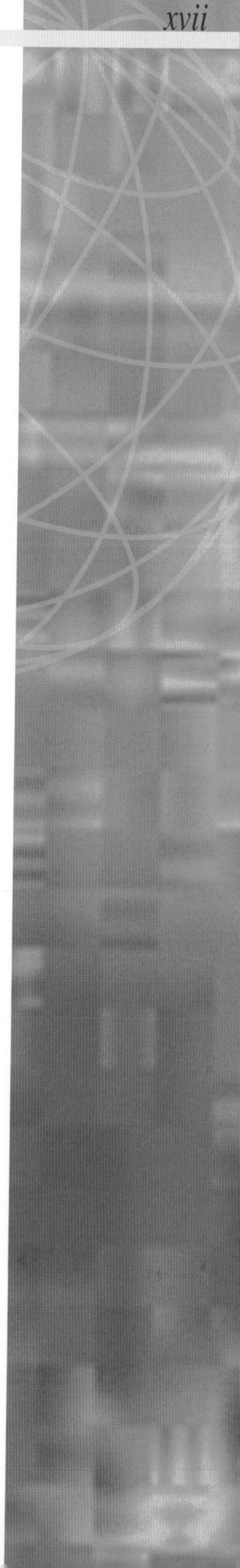

INNOVATION IN FAMILY FARMING

1. Innovation and family farming

Family farms and the challenges for world agriculture

Family farms are key to ensuring long-term global food security. To feed a growing population and eradicate poverty and hunger, family farms must be encouraged to innovate more and become more productive while also preserving natural resources and the environment.

Demand for food and agricultural products is increasing because the world's population is growing – to a projected 9.6 billion people in 2050 – and incomes are rising in much of the developing world. To satisfy added consumer demand, by 2050 global food production will have to increase by 60 percent from its 2005–2007 levels (Alexandratos and Bruinsma, 2012). However, producing this extra food will place additional stress on land, water and biodiversity, which are already scarce and showing worrying signs of degradation. In addition, climate change is likely to make it even more difficult to produce more food, and agriculture itself is a major source of greenhouse gas emissions. Meanwhile, in spite of remarkable advances in poverty reduction in many countries, significant levels of poverty remain in large parts of the developing world, especially in rural areas.

Family farms are central to meeting all of these challenges. More than nine out of ten farms in the world are family farms, making them the dominant form of farming in most countries.[1] The vast majority of farms in the world are smaller than 2 hectares. In low- and lower-middle income countries, farms smaller than 5 hectares manage the majority of agricultural land and produce a substantial portion of food.

However, many of these small and medium-sized farms have limited access to resources and low levels of productivity. If they are to help meet additional demand for food, preserve natural resources and fight poverty, they will need not only to grow more, but also to do so sustainably. In rural areas with high levels of poverty, enhancing the agricultural productivity of poor farmers can contribute dramatically to poverty alleviation and the reduction of undernourishment and malnutrition. According to the World Bank, gross domestic product (GDP) growth originating in agriculture raises the incomes of the poorest households by at least 2.5 times as much as growth in other sectors does (World Bank, 2007c).

Small family farms will not be able to increase their productivity and sustainability unless they are prepared to innovate and are supported in doing so. Given the critical importance of family farming to food security, natural resource preservation and poverty reduction, promoting innovation in family farming should be a priority for politicians and policymakers. The broad participation and involvement of farmers – including smallholders, women and disadvantaged or marginalized groups – will be essential.

[1] See Chapter 2 for a discussion of the concept of family farming.

Sustainable productivity growth in family farming occurs via two fundamental pathways (Table 1): (i) by developing, adapting and applying new technologies and practices for agricultural production and farm management; and (ii) by increasing and accelerating the adoption and application of existing technologies and practices. The first pathway expands the potential for more productive use of resources by pushing out the frontier of production possibilities. The second allows farmers to achieve more of this potential growth by moving towards the existing production possibilities frontier. The two pathways are not mutually exclusive but will generally be followed at the same time and reinforce each other. Both pathways are central to innovation in family farming and can be promoted through various instruments discussed in this report.

Concerning the first pathway, for millennia, farmers have experimented, adapted and innovated to improve their farming systems. More recently, such farmer-led innovation has been supplemented by formal scientific research, which has dramatically expanded the production possibilities frontier in agriculture, permitting large increases in agricultural productivity and output over past decades. Both farmer-led improvements and scientific research are important, and combining them can help ensure that agricultural research supports innovation among family farms.

The second pathway allows farmers to apply existing technologies and introduce more productive and sustainable practices. This pathway can be promoted by addressing some of the constraints that farmers face in introducing improved practices (e.g., limited access to finance, risk, insecure property and tenure rights), and providing incentives for adopting more sustainable practices. Effective extension and rural advisory services are fundamental for disseminating and sharing information about improved practices. Capacity to innovate can be promoted more widely through training and education to facilitate the formation of farmers' and local community groups (e.g., farmers' organizations) and the creation of an enabling environment for innovation.

Family farms and the agricultural innovation system

Farmers can innovate in different ways. Change can involve farm products (e.g., new types of crop or high-yielding varieties), production processes (e.g., zero-tillage or different crop rotations) and/or farm organization and management (e.g., new business models or ways of interacting with value chains, increasing storage capacity). Innovations in these different areas often occur concurrently.

Innovation can have different consequences. It may allow farmers to produce more with the resources and inputs they already have and to reduce their costs of production. It can allow them to expand, change or diversify their marketable output, increasing the profitability of their farms. It may also allow them to free up

TABLE 1
Pathways and instruments for sustainable productivity growth in agriculture

PATHWAY	TYPES OF INSTRUMENTS	DISCUSSED IN THE REPORT
Developing, adapting and applying new technologies and practices	Farmer-led improvements in technologies and practices Formal scientific research and development Combining farmer-led improvements and formal scientific research and development	Chapter 4
Accelerating and increasing adoption of existing technologies and practices	Addressing economic constraints to adoption of technologies and practices	Chapter 3
	Extension and advisory services (public and private) Promotion of innovation capacity	Chapter 5
	Individual (education, training) Collective (including producer organizations and cooperatives) Enabling environment for innovation (including linkages and networks)	Chapter 6

Source: FAO.

resources (e.g., labour) for use in other economic activities. Innovation can enhance the sustainability of production and/or the provision of important ecosystem services, both of which are more important than ever as natural resources become more constrained and more degraded.

There are many definitions of innovation in academic literature. Innovation in an economic context was first defined by Schumpeter (1939) as the introduction of a new production method, new inputs into a production system, a new good or a new attribute of an existing good, or a new organizational structure.[2] He clearly distinguished innovation from invention: "Innovation is possible without anything we should identify as invention, and invention does not necessarily induce innovation" (Schumpeter, 1939). Hayami and Ruttan (1971) elaborated the concept of induced technological innovation in agriculture (Box 1).

The Organisation for Economic Co-operation and Development (OECD) and Eurostat (2005) define innovation as "the implementation of a new or significantly improved product (good or service), or process, a new marketing method, or a new organizational method in business practices, workplace organization or external relations", which clearly mirrors Schumpeter's earlier definition. According to the World Bank (2010b), innovation "means technologies or practices that are new to a given society. They are not necessarily new in absolute terms, but they are being diffused in that economy or society. This point is important: what is not disseminated and used is not an innovation." This definition emphasizes that the recombination and use of existing knowledge is innovation. The World Bank (2010) also mentions the social benefits of innovation: "Innovation, which is often about finding new solutions to existing problems, should ultimately benefit many people, including the poorest."

A working definition elaborated by FAO and specific to the agricultural context focuses on the impact of innovation on food security, sustainability and development outcomes: "Agricultural innovation is the process whereby individuals or organizations bring existing or new products, processes and forms of organization into social and economic use to increase effectiveness, competitiveness, resilience to shocks or environmental sustainability, thereby contributing to achieve food and nutrition security, economic development and sustainable natural resource management" (FAO, 2012a).

These definitions characterize innovation as a process rather than a discrete event, and see it as fundamentally creative and geared towards solving problems. Innovation may not necessarily involve completely new knowledge or products: using existing inputs in new ways is also innovative.

Innovation is a complex process in which the different pathways and related instruments (Table 1) come into play simultaneously. Innovation in agriculture involves multiple actors such as farmers, producers' organizations and cooperatives, private companies in supply and value chains, extension services and national research organizations. Previously, the main focus of innovation was research as a means of generating technologies and knowledge, and extension as a means of disseminating the results of research. Recently, increasing attention has also been given to other sources of innovation. Potential benefits can be fully realized only if technologies and knowledge reflect real demand and are applied in combination with the ideas, practices and experience of farmers themselves.

Increasingly therefore, innovation is perceived as taking place within a network of actors – individuals and organizations – that fosters interaction and learning. The *innovation system* has gained prominence as an analytical concept that comprises the different sources and avenues of innovation and the relationships among the different actors involved in innovation processes. Since 2006, the World Bank, among others, has promoted this concept as a tool for enhancing agricultural innovation beyond the strengthening of research systems (World Bank, 2006). The World Bank defines the innovation system as a "network of organizations, enterprises and individuals focused on bringing new products, new processes and new forms of organization

[2] As cited in Phillips *et al.*, 2013.

into economic use, together with the institutions and policies that affect their behaviour and performance" (World Bank, 2008b). The innovation system concept recognizes the importance of technology transfer but also considers the social and institutional factors that establish linkages and networks among the various actors involved.

There is need to design an agricultural innovation system that meets the challenges of today, recognizes the importance of family farmers, and supports these farmers in innovating and achieving sustainable productivity increases. The challenges facing world agriculture are much more complex than they were in the 1940s and 1950s, when the institutions that gave rise

BOX 1

Induced technological innovation in agriculture

In their seminal work, *Agricultural development. An international perspective*, Hayami and Ruttan (1971) discuss the multiple paths of technical change available to societies. Different societies and farmers in different locations face different constraints to agricultural development. In some instances, land scarcity may be the most serious limiting factor, which can be addressed through advances in biological technology; in others cases, labour scarcity may be the most serious constraint, to which mechanical technologies may present the best response. Countries' achievement of growth in agricultural productivity and production depends on the ability to choose a pathway of technical change that relieves the constraints imposed by their respective resource endowments.

Hayami and Ruttan describe induced innovation in agriculture as a process in which technical change responds dynamically at different levels to changes in resource endowments and growth in demand. Induced technological innovation at the farm level occurs when farmers adapt their production methods to changes in demand and in the relative scarcity and prices of the main factors of production, such as land and labour. Such changes in relative prices may induce farmers to search for technical alternatives. Perceptive research scientists and administrators may then be induced to make available new technical possibilities and inputs that allow farmers to substitute factors that are less scarce for those that have become scarcer. This response by the research community represents a critical link in the process of induced innovation. The link is likely to be more effective when farmers are organized into politically effective organizations and associations. However, the authors do not argue that all technical change is induced; technical change can result from independent progress in science and technology.

At a different level, according to Hayami and Ruttan, technical change and changes in factor endowments and product demand may also lead to – or induce – institutional changes, such as the emergence of or change in institutionalized research at the national or international level, and changes in property right regimes or market institutions. Here too, collective action is important in bringing about these induced institutional changes. Cultural endowments can also have a powerful influence on institutional innovation, making some innovations easier to establish in some societies than others.

Hayami and Ruttan view the process of induced innovation as one in which resource endowments, technology, institutions and cultural endowments interact and influence each other in a dynamic process of development. The agricultural innovation system can therefore be seen as contributing to the effectiveness of these linkages and facilitating the adoption of a process of productivity growth and broader development that responds to the resource and institutional constraints facing individual countries at different stages of their development.

to the Green Revolution – the first major wave of organized agricultural innovation – were created. Since then, many of these institutions – international agricultural foundations and research centres, national agricultural research and extension systems, state marketing boards, cooperative producer groups, and the broader enabling environment for innovation – have been disbanded, underfunded or allowed to drift from their central mission. Today, new actors have entered the scene, including private agricultural research and technology companies and a range of civil society providers of agricultural advice, creating a much more complex institutional context for agricultural innovation.

Increasing urbanization, globalization and demand for high-value products have also dramatically changed the global context for agriculture. Value chains are becoming more important, and pressure is mounting to preserve the natural resource base for agriculture, especially given advancing climate change. Innovation systems must allow family farmers to meet these different challenges. There is need to:

- design innovation systems that are responsive to farmers' needs and demands by:
  - making farmers protagonists in, rather than mere recipients of, agricultural innovation;
  - supporting the development of organizations, linkages and networks involving family farms;
- promote collective and individual capacity to innovate;
- recognize the diversity of family farms and of the demands and needs of different household members and value chains, which call for tailored policies and targeted reforms.

This report focuses on promoting agricultural innovation among family farms. However, it is important to recognize the limitations of such innovation for rural development and poverty alleviation. Promoting agricultural innovation among family farms is a central part of an agriculture-based poverty alleviation strategy, but additional options are needed for many small family farms. These farms, especially the smaller ones, often already have diversified livelihoods and sources of income; agriculture cannot be their sole or even their main source of income if they are to escape poverty. To alleviate rural poverty while avoiding socially undesirable urbanization rates, many small family farms must be able to rely on other sources of income to supplement, and sometimes replace, the income derived from farming. Vibrant rural economies and a range of other policy instruments are needed (e.g., social protection and rural development), which are beyond the scope of this report.

Structure of the report

Chapter 2 discusses family farming, its prevalence, role and capacity to innovate. Chapter 3 addresses the challenge of sustainable productivity growth and some of the barriers and disincentives that prevent farmers from adopting more productive and sustainable practices. Chapter 4 looks at trends and issues in agricultural research and the challenge of ensuring that research responds to the needs of family farms. Chapter 5 deals with extension and advisory services, and how to make them more inclusive and responsive. Chapter 6 discusses how to promote innovation capacity more broadly. Chapter 7 summarizes the report's main conclusions.

2. Family farming

At least 90 percent of the world's farms are family farms according to the most commonly used definitions.[3] Family farms represent the dominant form of agriculture in most countries. They range in size from tiny, subsistence holdings to large-scale, commercial enterprises, and they produce a vast range of food and cash crops in all kinds of agro-ecological conditions. However, the enormous heterogeneity of family farms means that general policy prescriptions are unlikely to be relevant for the whole category. It is necessary to look at the different characteristics of farms within the broad category of family farming. This chapter briefly reviews the state of family farming in the world, focusing on smaller family farms.

What is a family farm?

Although there is no universal agreement on what constitutes a family farm, many definitions refer to factors related to ownership and management, labour use, and physical or economic size. In a survey of 36 definitions of family farm, nearly all specify that the farm is owned, operated and/or managed at least partly by a member of the household; many specify a minimum share of labour contributed by the owner and his/her family; many set upper limits on the land area or sales of the farm; and some also set upper limits on the share of household income derived from non-farm activities (Garner and de la O Campos, 2014). Even this broad range of definitions does not capture the diversity of concepts incorporated under the term (Box 2). At least one country is reportedly using the conceptual definition of a family farm to promote the consolidation of very small production units into larger more economically viable farms (*News China Magazine*, 2013).

How prevalent are family farms?

Based on the most common elements of definitions of family farms, and information obtained from several rounds of national agricultural censuses, FAO made a broad assessment of the number of farms in the world and the worldwide prevalence of family farms for this report. The best available proxy measure for farms reported in the censuses is the agricultural holding.[4] The total number of agricultural holdings in the world was estimated at about 570 million.

As noted in the previous section, most definitions of a family farm require that the farm be partially or entirely owned, operated and/or managed by an individual and her/his relatives. Information on the legal status of the agricultural holder[5] can be found in a number of agricultural censuses. In almost all the countries where this information is available,[6] for more than 90 percent of farms (and often close to 100 percent)

[3] Unless otherwise noted, the analysis in the first two sections of this chapter is based on a background paper by Lowder, Skoet and Singh (2014). Data used are from several rounds of the FAO World Programme for the Census of Agriculture, especially FAO (2013a) and FAO (2001).

[4] FAO's theoretical definition of an agricultural holding is "an economic unit of agricultural production under single management comprising all livestock kept and all land used fully or partly for agricultural production purposes, without regard to title, legal form, or size. Single management may be exercised by an individual or household, jointly by two or more individuals or households, by a clan or tribe, or by a juridical person such as a corporation, cooperative or government agency" (FAO, 2005a). FAO encourages countries to use an operational definition based on this theoretical definition when carrying out their agricultural censuses.

[5] FAO defines the agricultural holder as "the civil or juridical person who makes the major decisions regarding resource use and exercises management control over the agricultural holding operation. The agricultural holder has technical and economic responsibility for the holding and may undertake all responsibilities directly, or delegate responsibilities related to day-to-day work management to a hired manager" (FAO, 2005a).

[6] 52 countries report data on the legal status of the agricultural holder.

BOX 2
The definition of family farming for the International Year of Family Farming

The International Steering Committee for the International Year of Family Farming, celebrated in 2014, developed the following conceptual definition of family farming:

> *Family Farming (which includes all family-based agricultural activities) is a means of organizing agricultural, forestry, fisheries, pastoral and aquaculture production which is managed and operated by a family and predominantly reliant on family labour, including both women's and men's. The family and the farm are linked, co-evolve and combine economic, environmental, social and cultural functions.*

Source: FAO, 2013b.

the agricultural holder is an individual, a group of individuals or a household, with or without a formal contract. In the remaining cases, the holder is an entity such as a corporation, a cooperative or a public or religious institution.

Several definitions of family farm also require that the family supply most of the labour on the farm. Relatively few agricultural censuses provide information on labour supply; those that do report that, on average, about half the family members are engaged in part- or full-time labour on the homestead.[7] Conversely, the average number of permanent hired workers on family farms is very small (well below one per farm) in nearly all countries where such information is available.[8] The average ratio of family members working on the farm to permanent hired farm workers is 20 to 1.[9]

The available evidence thus suggests that family farms, as commonly defined, account for more than 90 percent of farms in most countries. With about 570 million farms in the world, the total number of family farms consequently exceeds 500 million.[10]

Family farms occupy large tracts of the world's farmland and contribute substantially to the world's food supply. However, family farms are likely to own less than 90 percent of total farmland, because non-family farms tend to be larger. Lack of data makes it impossible to assess the exact share at the global level, but in a sample of 30 countries[11] an average of about 75 percent of farmland is held by households or individuals.[12] Based on the share of land held by family farms and the value of food production in each country, it is estimated that family farms produce more than 80 percent of the food in these countries.[13] Using a different methodological approach, Graeub *et al.* (forthcoming) also concluded that there are more than 500 million family farms in the world and that they supply most of the world's food production.

[7] 15 countries report data on the share of household labour engaged in farming.
[8] 65 countries report data on the number of permanent hired workers.
[9] 31 countries report data on the numbers of both family members and permanent hired workers working on farms.

[10] Because of data limitations, the figure for family farms worldwide should be considered an approximation. Current agricultural censuses are not available for many countries where farm fragmentation is taking place, so the total number of farms may exceed 570 million. In addition, in almost all countries for which data are available, 90 percent represents a conservative estimate of the share of family farms in the total. On the other hand, agricultural censuses do not provide data on seasonal workers, who are often an important source of labour for farms. Accurate data on the use of seasonal labour might lead to lower estimates of the share of family farms in several countries, depending on the threshold used for the share of non-family labour in the family farming definition.
[11] These countries contribute 35 percent of the world's food production in value.
[12] The unweighted average share is 73 percent and the weighted average is 77 percent.
[13] This estimate is based on the share of land held by individuals or households (farming families) in each of the 30 countries. In each country, it is assumed that the share of food produced by family farms corresponds to their share of land. This allows estimation of the value (in international dollars) of food produced by family farms in each country based on the total value of food produced in the country. Adding the values of food produced by family farms in each of the countries and dividing by the total value of food produced in all 30 countries, results in a share of 79 percent. However, family farms tend to be smaller than non-family farms, and (as discussed in the following section) small farms in individual countries tend to have higher yields per hectare than larger farms. The share of food produced by family farms is therefore likely to be larger than 80 percent, although the exact share cannot be quantified.

Distribution of farms around the world

Of the world's 570 million farms, almost 75 percent are located in Asia (Figure 1): China and India account for 59 percent (35 percent and 24 percent respectively); 9 percent are in other countries of East Asia and the Pacific; and 6 percent are in other South Asian countries. Only 9 percent of the world's farms are located in sub-Saharan Africa, 7 percent are in Europe and Central Asia, 4 percent in Latin America and the Caribbean, and 4 percent in high-income countries. About 47 percent of farms are in upper-middle-income countries, including China, and 36 percent in lower-middle-income countries, including India.

The vast majority of these farms are small by any definition. Small farms are frequently defined in terms of physical size, and farms are often considered small when they are less than 1 or 2 hectares. According to agricultural census data from a large sample of countries, 72 percent of farms are less than 1 hectare, and 12 percent are between 1 and 2 hectares (Figure 1).[14] This is similar to the distribution of farm sizes found by the High Level Panel of Experts on Food Security and Nutrition of the Committee on World Food Security[15] (HLPE, 2013). Assuming this distribution to be representative of farm sizes throughout the world, it can be estimated that there are 400 million farms of less than 1 hectare, and 475 million of less than 2 hectares.[16]

It is not possible to estimate global or regional numbers of farms in size categories below 1 hectare because of the lack of data for a sufficient number of countries. However, in many countries, farms that are significantly smaller than 1 hectare – such as those below 0.5 hectares – constitute a significant share of the total. In India,[17] for example, 47 percent of farms are smaller than 0.5 hectares; in Bangladesh,[18] 15 percent are. In Africa, the shares of farms of less than 0.5 hectares are as high as 57 percent in Rwanda[19] and 44 percent in Ethiopia,[20] but only 13 percent in the United Republic of Tanzania,[21] 11 percent in Senegal[22] and 10 percent in Mozambique.[23] In Latin America the shares are 6 percent in Brazil[24] and 2 percent in Venezuela.[25]

While farms of less than 2 hectares account for more than 80 percent of all farms at the global level, they occupy a far smaller share of the world's farmland. Agricultural census data suggest that farms of more than 50 hectares occupy two-thirds of the world's farmland, while farms of up to 2 hectares cover only about 12 percent (Figure 2).[26] However, these figures reflect the situation mainly in high-income and upper-middle-income countries, especially in Latin America. The situation is substantially different in low-income and lower-middle-income countries, where small farms (up to 2 hectares) occupy large shares of farmland (Figure 3), which become even larger if medium-sized farms up to 5 hectares are included. In lower-middle-income countries, farms of up to 2 hectares occupy more than 30 percent of the land and farms of up to 5 hectares about 60 percent. In low-income countries, farms up to 2 hectares occupy about 40 percent of the land and those up to 5 hectares about 70 percent. The shares of small farms in food production are likely to be even larger as evidence indicates that smaller farms tend to have higher output per hectare than larger farms (see following section). In other words, at least in low- and lower-middle-income nations, small and medium-sized family farms make a crucial contribution to food security.

The distribution of farm sizes across countries and over time depends on complex factors such as history, institutions, economic development, the development of the non-farm sector, land and labour

[14] The sample includes 111 countries.
[15] The HLPE report examined results from the 2000 round of agricultural censuses, with 81 countries in the sample.
[16] The world's 570 million farms multiplied by 72 percent and 84 percent respectively.
[17] Data from the Government of India (2012).
[18] Data from the Government of Bangladesh (2010).
[19] Data from the Government of Rwanda (2010).
[20] Data from the Government of Ethiopia (2008).
[21] Data from the Government of the United Republic of Tanzania (2010).
[22] Data from the Government of Senegal (2000).
[23] Data from the Government of Mozambique (2011).
[24] Data from the Government of Brazil (2009).
[25] Data from the Government of Venezuela (2008).
[26] These figures are derived from a sample of 106 countries that are, by most measures, fairly representative of farms around the world; together they represent about 450 million, or 80 percent, of the world's farms and account for 85 percent of the world's population active in agriculture, and 60 percent of agricultural land (FAO, 2014b).

FIGURE 1
Shares of the world's farms, by region, income group and size

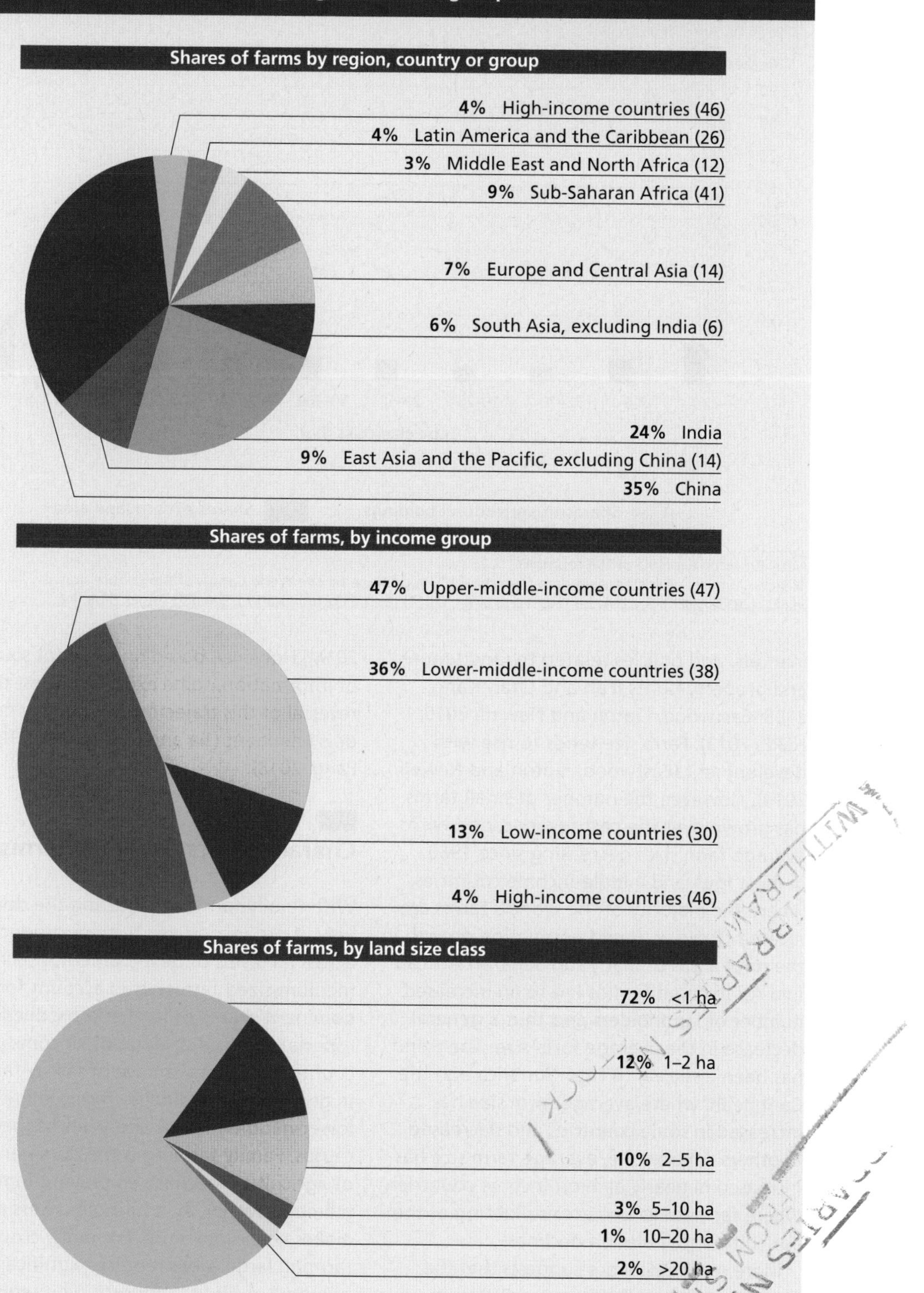

Note: The first two panels are based on a sample of 161 countries, which account for almost 570 million farms; the number of countries is shown in parentheses. The third panel shows farms by farm size covering a total of about 460 million farms in 111 countries. Countries included are those for which data were available from the World Census of Agriculture and for which the World Bank (2012a) provided regional and income groupings. All figures are rounded.
Source: Authors' compilation using data from FAO (2013a; 2001) and other sources from the FAO Programme for the World Census of Agriculture. See Lowder, Skoet and Singh (2014) for full documentation. See also Annex tables A1 and A2.

FIGURE 2
Distribution of farms and farmland area worldwide, by land size class

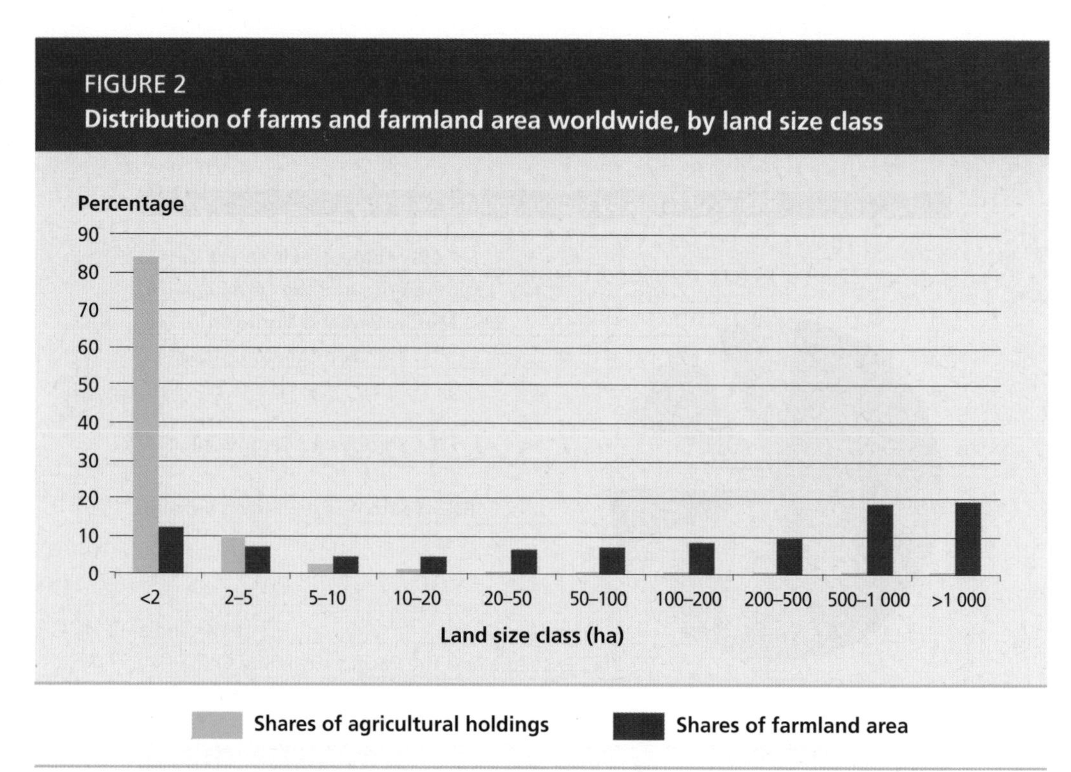

Note: Based on a sample of 106 countries.
Sources: Authors' estimates using data from the FAO Programme for the World Census of Agriculture shown in FAO (2013a; 2001). See Lowder, Skoet and Singh (2014) for full documentation. See also Annex table A2.

markets, and policies related to land tenure and property rights (Fan and Chan-Kang, 2005; Eastwood, Lipton and Newell, 2010; HLPE, 2013). Farm size tends to rise with development (Eastwood, Lipton and Newell, 2010). However, the number of small farms has grown over the past few decades, with average farm sizes decreasing since 1960 in most low- and middle-income countries, where the majority of the world's farms are located (Table 2). Rapid population growth in the rural areas of many sub-Saharan African and Asian countries has led to an increased number of landholders and thus a general decrease in the average farm size. The trend has been less clear in Latin America and the Caribbean, where average farm size has increased in some countries and decreased in others. Meanwhile, average farm size has increased in nearly all high-income countries, where farms have been consolidating as the agricultural population declines.

More recent evidence suggests that the trend towards smaller farms continues in Africa, but that consolidation may have begun in Asia (Masters *et al.*, 2013). In China, agricultural censuses show a decrease in average farm size from 0.7 hectares in 2000 to 0.6 hectares in 2010 (Lowder, Skoet and Singh, 2014). However, based on different sources of information, some experts suggest that a reversal of this trajectory has already occurred or is imminent (Jia and Huang, 2013; Nie and Fang, 2013).

Characteristics of family farms

With family farms constituting the dominant way of organizing agricultural production across all levels of development, small and medium-sized farms often account for the dominant shares of land and production, especially in low- and middle-income countries. The prevalence of family farms in general and of smaller farms in low- and lower-middle-income countries has several causes. Family farming is the dominant form of agriculture because employing family members rather than hiring workers usually makes economic sense. For many crops, farming large areas requires significant numbers of hired labourers, who require supervision. Supervision costs often outweigh any benefits from economies of scale, making family farms the best solution in many agricultural contexts. The size of family farms is also often limited to what the family

FIGURE 3
Distribution of farms and farmland area, by land size class and income group

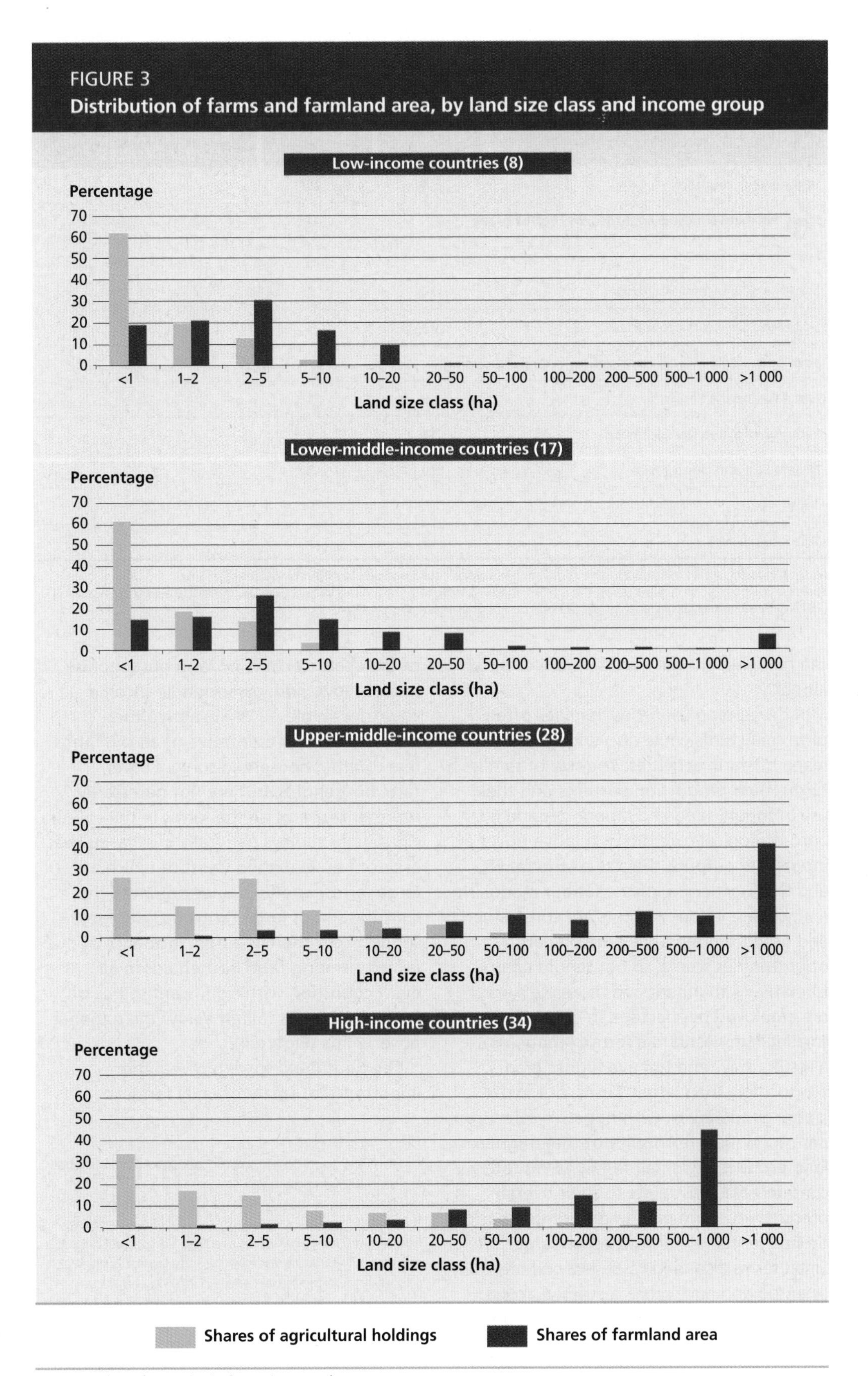

Note: Number of countries is shown in parentheses.
Sources: Authors' compilation using data from the FAO Programme for the World Census of Agriculture shown in FAO (2013a) and FAO (2001). See Lowder, Skoet and Singh (2014), for full documentation. See also Annex tables A1 and A2.

TABLE 2
Number of countries exhibiting a decrease or increase in the average size of agricultural holdings, 1960–2000, by income and regional groupings

COUNTRY GROUPING	DECREASE	INCREASE	NEITHER CLEAR INCREASE NOR DECREASE
High-income countries	**6**	**25**	**4**
Low- and middle-income countries, by income group			
Low-income countries	12	2	1
Lower-middle-income countries	24	2	0
Upper-middle-income countries	19	5	1
Low- and middle-income countries, by regional grouping			
East Asia and the Pacific	9	1	0
Latin America and the Caribbean	18	7	2
Middle East and North Africa	10	0	0
South Asia	5	0	0
Sub-Saharan Africa	15	3	1

Note: A few countries included in the regional groupings could not be classified by income groups.
Sources: Authors' compilation using data from the FAO Programme for the World Census of Agriculture shown in FAO (2013a). See Lowder, Skoet and Singh (2014) for full documentation.

can manage without excessive use of hired labour.

In developing countries, families often farm small plots while also engaging in many off-farm activities. The size of family farms, their production patterns and their use of inputs, land and labour depend on agro-ecological conditions, relative prices of inputs and outputs, the size of the family, and the functioning of the labour market. In many cases, labour markets are constrained and other remunerative employment opportunities scarce, so household labour is relatively abundant and more workers are employed per hectare. In general, smaller farms tend to overuse labour. As a result, they tend to have higher land productivity than larger farms, but lower labour productivity, with negative effects on per capita income. In spite of their higher land productivity, small family farms face considerable constraints to their overall productivity. Farm equipment is more basic on smaller family farms than on larger ones. Small farms also tend to be less commercially oriented and have more restricted access to markets for inputs, outputs, credit and labour.

Research conducted by FAO (see also Rapsomanikis, 2014) used household income and expenditure surveys to examine some of the characteristics of farm households[27] in eight low- and lower-middle-income countries (Table 3). While agricultural censuses are representative of all the farms in a country, household surveys cover farm households, but are not necessarily representative of all the farms in the country. Household surveys generally miss farms that are not family-owned (most of which are large farms) and thus underestimate the contribution of large farms.[28] The surveys suggest that there is a high incidence of poverty among farm households in all eight countries, with significant shares of farm households falling below the national poverty line (Figure 4).

The household surveys reveal the importance of smaller family farms to

[27] From this point on the words "household" and "family" are used interchangeably.
[28] For most countries, it is not possible to determine the extent to which larger farms are excluded from the household surveys based on available agricultural census reports. In Nicaragua, for example, the largest farm size cohort in the agricultural census is 200 ha and above (FAO, 2013a), which represents 30 percent of the country's farmland and averages about 475 ha per farm (see Annex table A2). This suggests that there are several farms larger than those described in the household survey data (in which farms were a maximum of 282 hectares) and that these larger farms contribute significantly to overall food and agricultural production.

TABLE 3
Number, average size and maximum size of household farms in surveys, by country

COUNTRY	NUMBER OF FARMS	AVERAGE FARM SIZE	MAXIMUM FARM SIZE
	(Thousands)	*(ha)*	*(ha)*
Bangladesh	14 950	0.4	2
Bolivia	680	1.5	151
Ethiopia	n.a.	1.9	19
Kenya	4 320	0.9	8.9
Nepal	3 260	0.9	17
Nicaragua	310	9.5	282
United Republic of Tanzania	4 700	1.5	21
Viet Nam	11 460	0.7	12

Note: n.a. = not applicable.
Source: FAO, 2014a.

FIGURE 4
Poverty headcount ratios for farm household populations

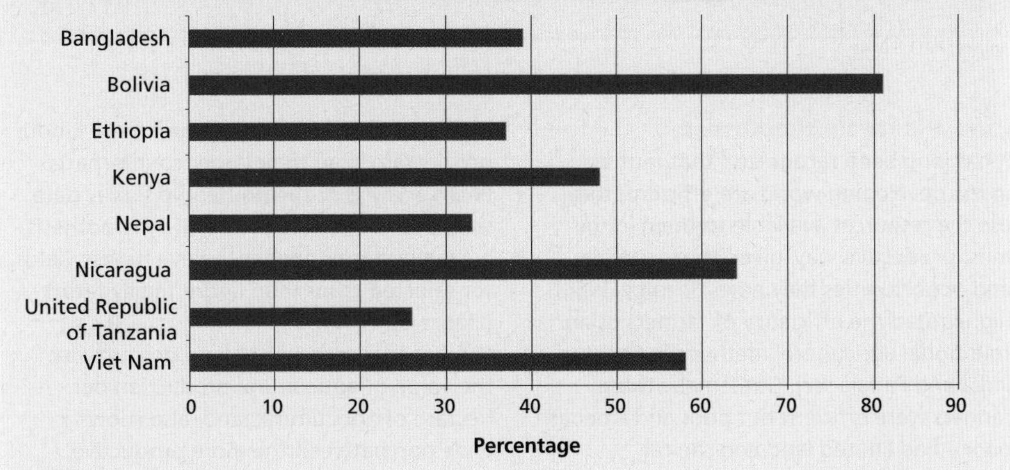

Bangladesh
Bolivia
Ethiopia
Kenya
Nepal
Nicaragua
United Republic of Tanzania
Viet Nam
0 10 20 30 40 50 60 70 80 90
Percentage

Notes: National poverty lines are used to calculate the poverty headcount ratio, which is the prevalence of poverty among the population living in farm households. Cross-country comparisons are not possible due to the use of country-specific poverty lines.
Source: Rapsomanikis, 2014.

food production. Although it does not indicate what share of national agricultural production is attributable to family farms, a sample of seven countries shows that the smallest 75 percent of family farms[29] are responsible for the greater part of food production by households (Figure 5).[30] As they use less than 50 percent of the total agricultural land operated by family farms, these smaller family farms have higher land productivity than do the larger ones.

[29] Throughout the rest of this chapter, farms are considered by size using the farmland quartile. Each quartile contains 25 percent of the farms in the country sample: the first quartile contains the smallest farms, and the fourth contains the largest. The 75 percent smallest farms are those in the first three quartiles.

[30] Their share of total national food production may be smaller, depending on the extent to which larger farms are excluded from the sample.

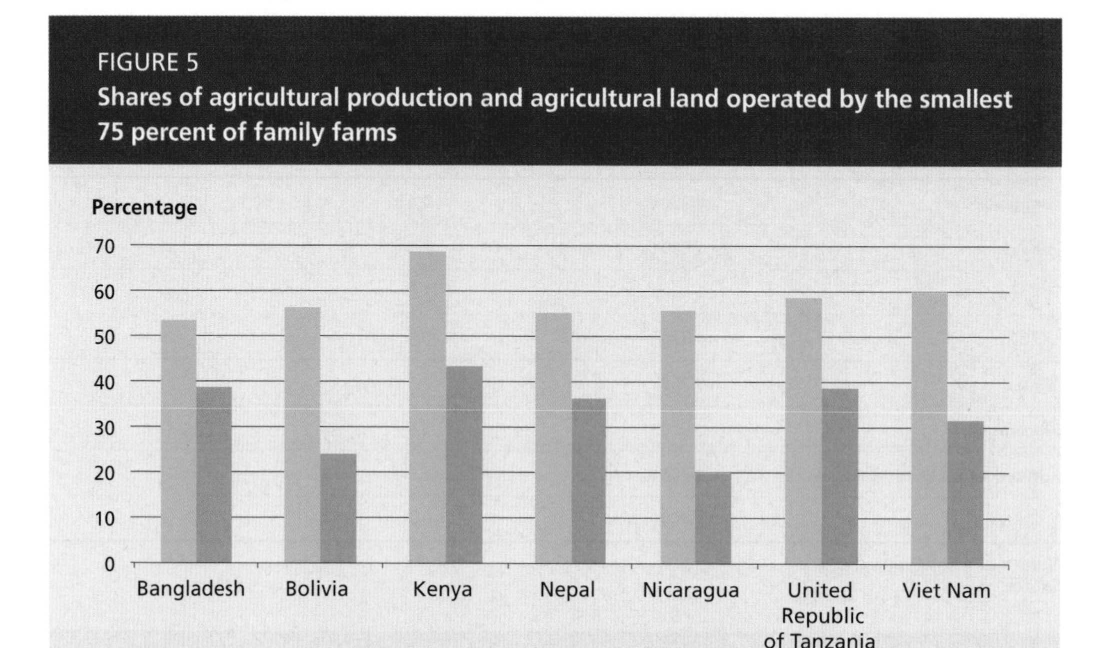

FIGURE 5
Shares of agricultural production and agricultural land operated by the smallest 75 percent of family farms

Source: FAO, 2014a.

Land and labour productivity

It has long been recognized that farmers in the developing world are efficient: they use the resources available to them in the most productive way, given the incentives and opportunities they have. Schultz (1964) highlighted the efficiency of farmers using traditional agricultural methods in Senapur, India and Panajachel, Guatemala: these farmers were efficient but poor and – being poor – had limited land and capital.

In more recent years, a large body of literature on land productivity by farm size has shown a phenomenon referred to as the "inverse productivity relationship", i.e. in a number of countries smaller farms have higher crop yields than do larger ones (Larson *et al.,* 2013; Barrett, Bellemare and Hou, 2010).[31] Larson *et al.* (2013) show that in each country in a sample of sub-Saharan African countries, smallholder maize farmers have higher land productivity but use more labour per hectare than their larger counterparts. FAO's analysis of the household survey data supports the inverse productivity hypothesis, as smaller farms appear to have higher yields for selected crops than larger family farms (Figure 6).

A broader measure of land productivity, the value of agricultural production per hectare of agricultural land, also shows a wide gap between the more productive, smaller family farms and the larger ones (Figure 7). With labour productivity, the situation is the reverse: in most of the sample countries, smaller family farms show far lower labour productivity than do larger farms. In short, smaller family farms have higher land productivity but lower labour productivity than larger family farms. Low labour productivity implies lower household incomes and consumption. The surveys show that households with smaller farms have lower incomes and consumption and substantially higher poverty rates than do households with larger farms (Rapsomanikis, 2014).

Low labour productivity often reflects an excessive use of farm labour – generally unpaid family labour – resulting from a

[31] The inverse productivity relationship refers to situations within countries and with comparable agro-ecological and socio-economic conditions. Both land and labour productivity are higher on large farms in high-income countries using advanced agricultural technologies than on small farms in low-income countries.

FIGURE 6
Selected crop yields, by farm size

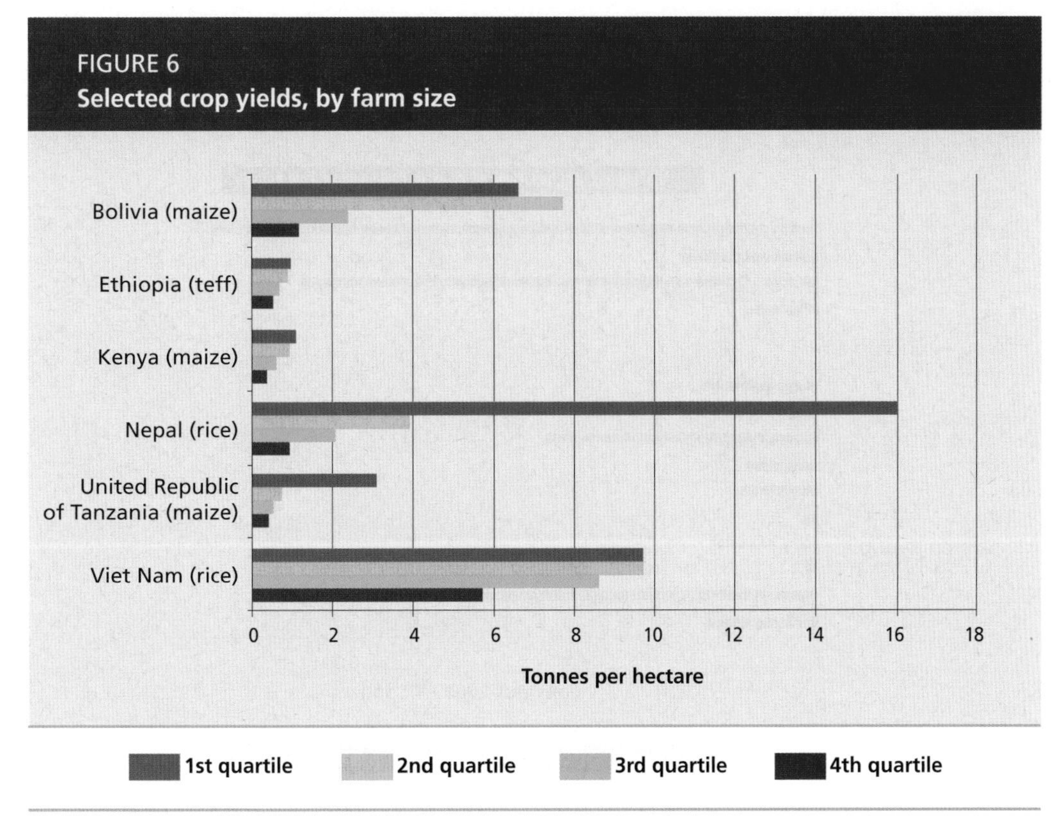

Source: FAO, 2014a.

scarcity of alternative sources of employment and income and a poorly functioning labour market. Karfakis, Ponzini and Rapsomanikis (2014) explore household survey data from Kenya and find that Kenyan maize farmers systematically overuse labour and underuse inputs such as seeds and fertilizer.[32] The overuse of labour is greater on smaller farms than larger farms, while the underuse of inputs is greater on larger farms. The authors theorize that these imbalances result from lack of access to natural resources, and the imperfect functioning of input, labour and land markets. In an analysis of nationwide data from Rwanda, Ali and Deininger (2014) find confirmation of the inverse productivity relationship and cite labour market imperfections as the key reason.

Multiple income sources

For most farming families, agriculture is only one of several sources of income (Rapsomanikis, 2014). Engaging in a wide range of off-farm activities represents both an attempt to make the best use of available household labour and a form of risk management. Smaller family farms tend to rely more on off-farm income than do larger ones, partly because their small plots usually yield insufficient incomes. Farming is more often the main source of revenue for larger farms (Figure 8). The share of income from farming increases with farm size in all eight countries in the household survey sample. In Bangladesh, for example, this share averages about 20 percent for the smallest farms (those in the first quartile) and about 65 percent for the largest (the fourth quartile).

Because of their reliance on multiple sources of income, smaller farms are more seriously affected than larger ones by a lack of adequate alternative employment opportunities and poor remuneration for any work that is available. For the smallest family farms, escaping poverty requires not only increasing farm labour productivity, but also the creation of non-farm employment opportunities through rural development, more efficient labour markets, and strengthening of the skills and capacities

[32] They overuse labour in that the value of the marginal output obtained by employing one additional unit of labour is less than the cost of this labour. In other words, farmers could earn more by using some of their farm labour in activities outside the farm.

FIGURE 7
Land and labour productivity, by farm size

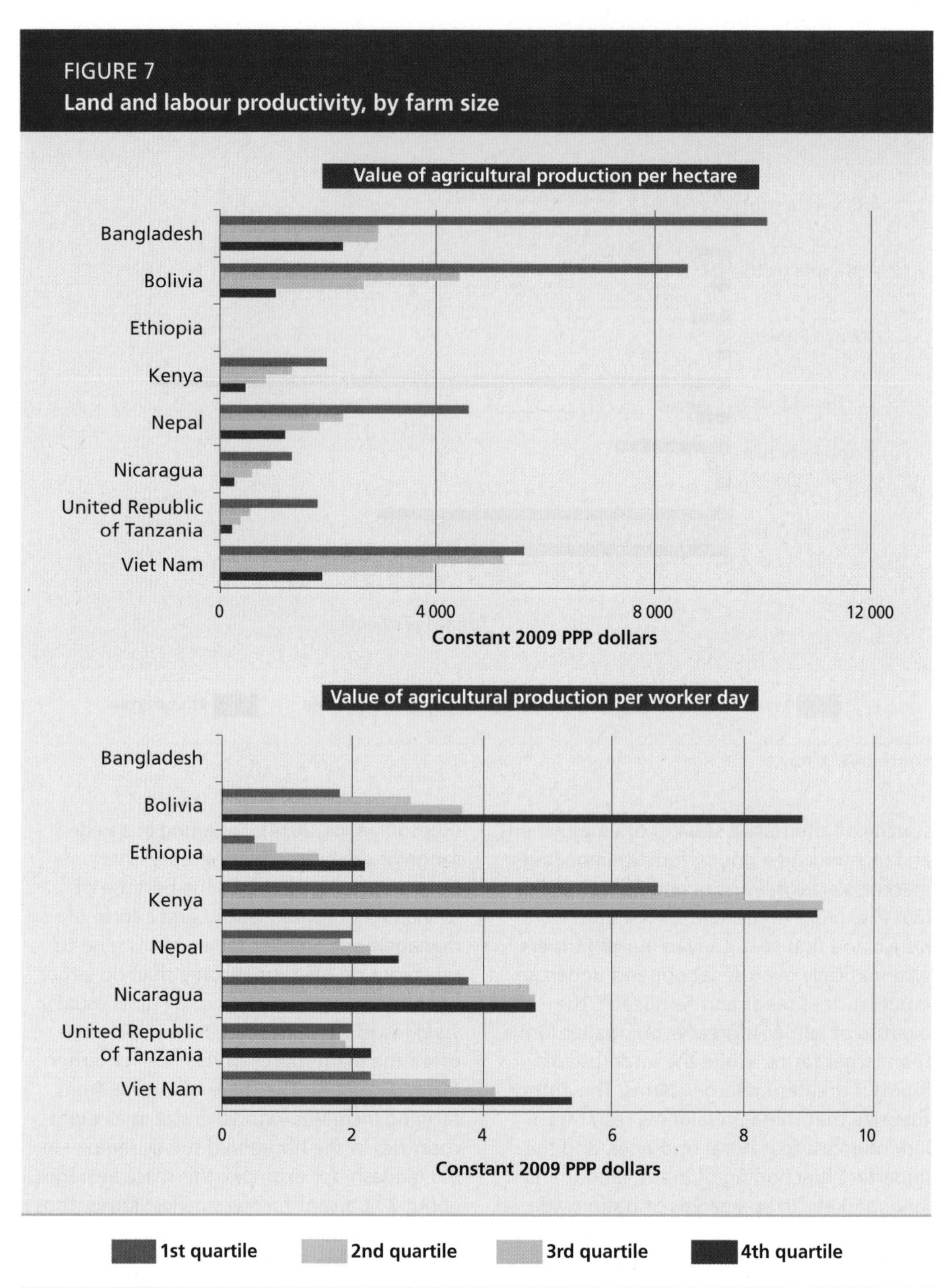

Notes: Land productivity is measured as the value of agricultural production (constant 2009 PPP dollars) per hectare of agricultural land. Labour productivity is the value of agricultural production (constant 2009 PPP dollars) per worker day, with workers including a measure of hired labour as well as household labour for all countries except Viet Nam, where no information was available on hired labour. The estimates of labour productivity are more appropriate for analysis by farm size within each country, rather than for cross-country analysis, because the method for estimating labour days varies from one survey to the next, based on the data available.
Source: FAO, 2014a.

of farm household members. Access to alternative sources of employment can allow farmers to diversify their sources of income and reduce their dependence on agriculture. It can also affect farm innovation, such as by stimulating the adoption of labour-saving technologies. Broader rural development and possibilities for economic diversification can therefore be major drivers of innovation in agriculture.

FIGURE 8
Average shares of household income, by source and farm size

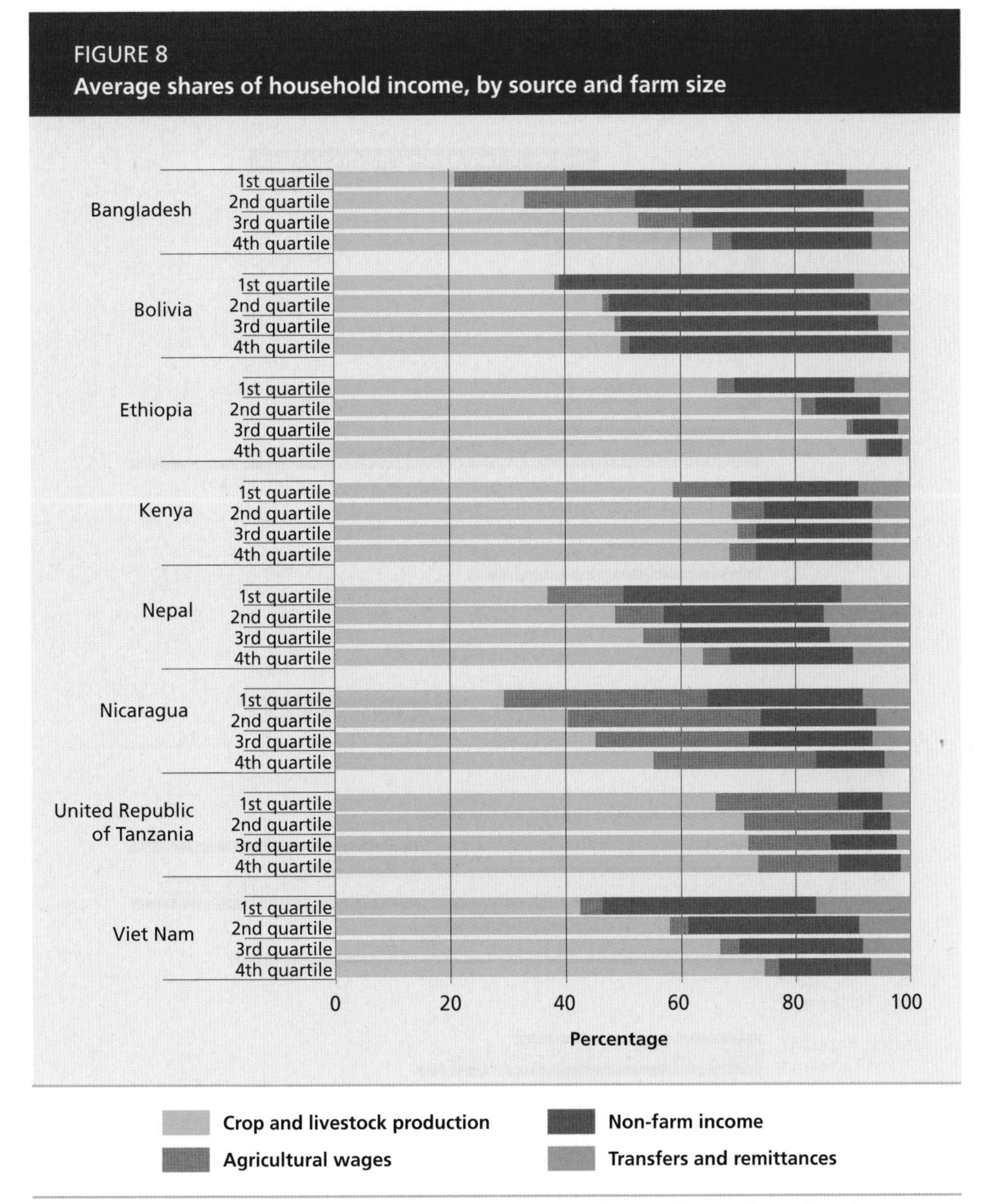

Note: Non-farm income includes wages for non-farm wage employment and income from non-farm self-employment; in other words, it is income earned through non-agricultural activities.
Source: FAO, 2014a.

Use of modern farming technology

Low labour productivity on the smallest farms may reflect not only the excessive amount of labour used, but also the farming technologies applied. In many of the countries considered, both large and small farms make limited use of mechanized technologies and improved seeds, but use is particularly limited on smaller farms (Figure 9). Although the low levels of mechanization reflect the abundance of family labour, there would also seem to be much scope for increasing agricultural productivity by promoting greater use of existing technologies and farming processes.

There are also major differences in the volumes of inputs used among countries. Rapsomanikis (2014) notes that the average quantity of fertilizer used on farms (regardless of farm size) in many of the countries in the household survey sample is far lower than that used in high-income

FIGURE 9
Shares of farms using selected modern farming technologies, by farm size

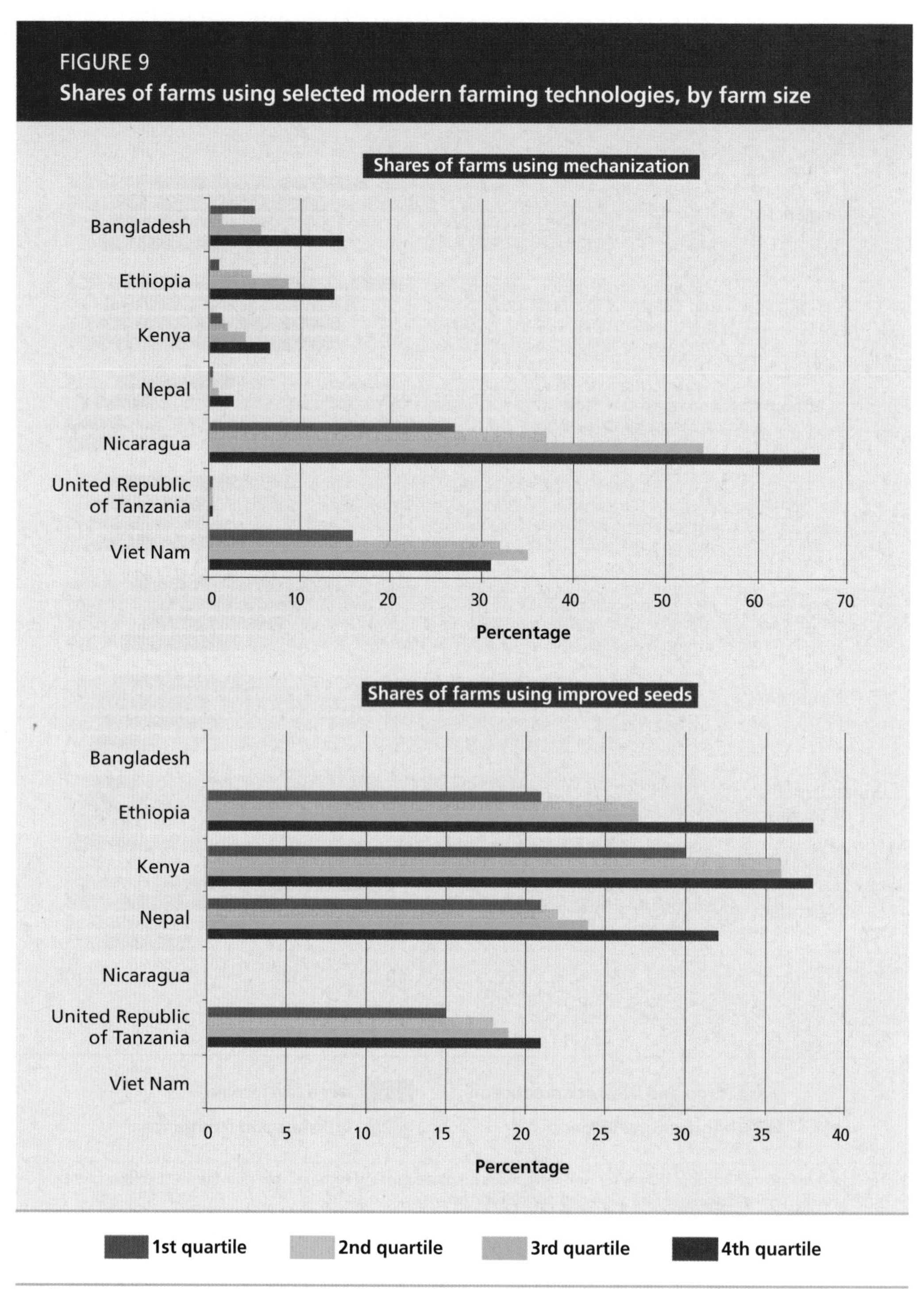

Source: FAO, 2014a

countries in Europe. However, in nearly all the eight sample countries, smaller farms use more seeds and fertilizer per hectare than larger farms (Figure 10). This is similar to the situation with regard to labour and reflects many factors, including economic choices and differences in farming systems and agro-ecological conditions. It suggests that smaller family farms strive to get as much as possible from their small plots by applying larger amounts of both labour and key inputs.

Access to markets

Many small family farms grow food for only their own consumption, but there is often scope for increasing their productivity

FIGURE 10
Intensity of seed and fertilizer use, by farm size

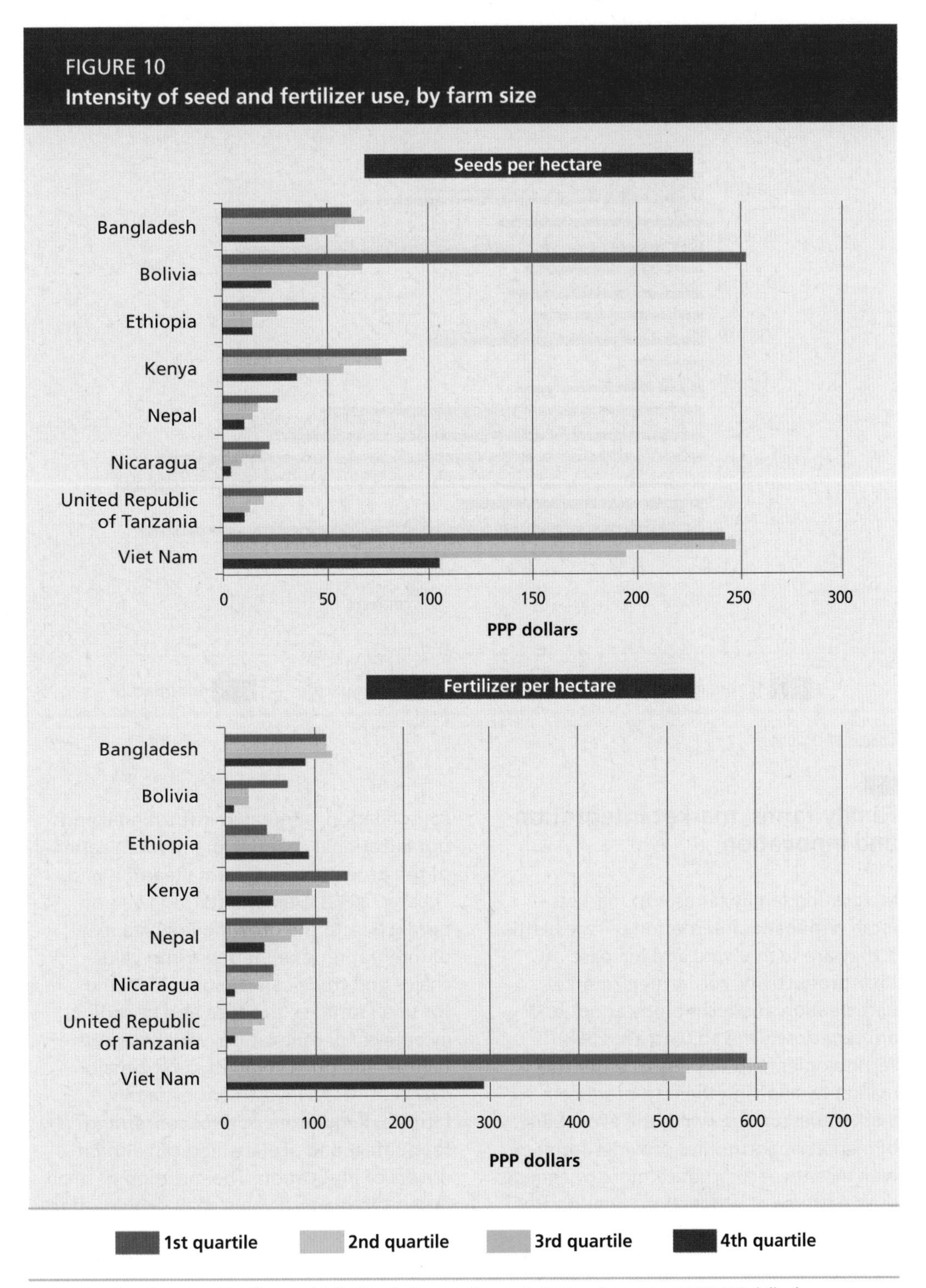

Note: Quantity of seed and fertilizer multiplied by their respective market prices (in constant 2009 PPP dollars).
Source: FAO, 2014a.

and output. For this to happen, it is crucial that small farms enter markets. Such market entry may involve greater specialization or improved marketing of the diversified product mixes that small farmers are often expert at producing. In most of the household survey countries, smaller farmers sell a smaller average share of their agricultural production than do larger farmers (Figure 11). To some extent, this reflects the greater availability of marketable surplus production on larger farms, but is likely also to reflect the choice of farm products (e.g. food crops versus cash crops).

FIGURE 11
Shares of agricultural production sold, by farm size

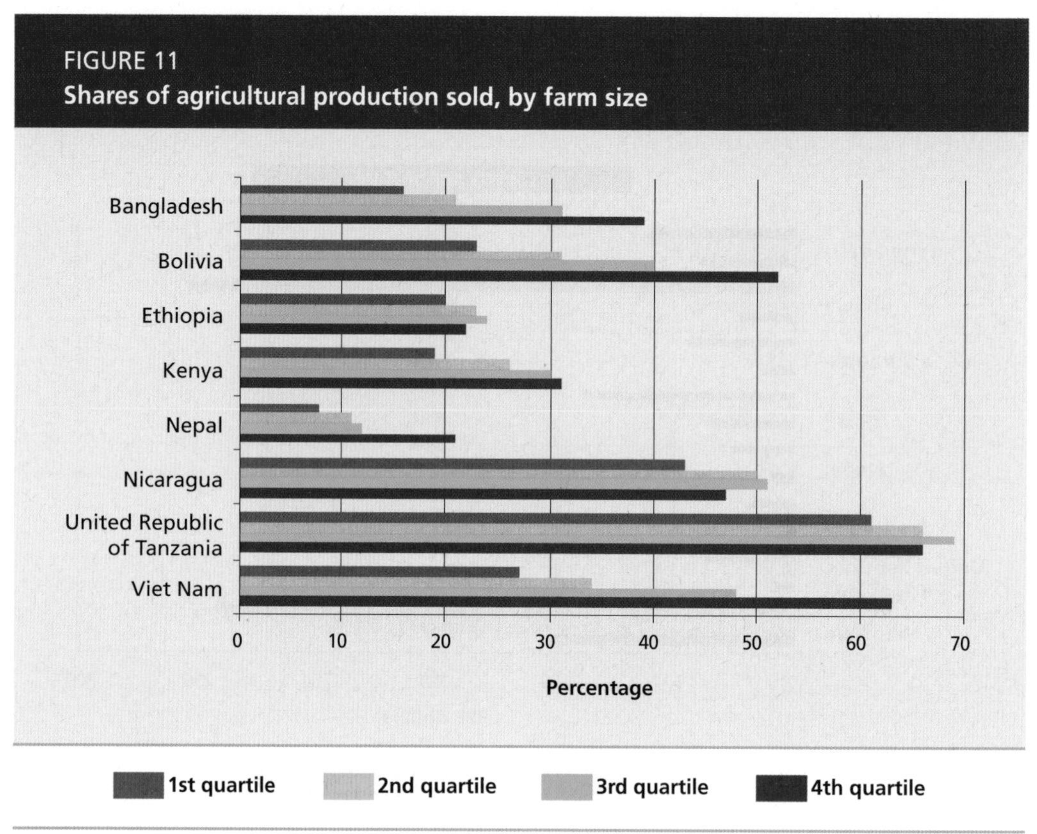

Source: FAO, 2014a.

Family farms, market integration and innovation

Integrating family farms into markets – local, national or international – is essential if they are to innovate and increase their productivity. For farmers, market participation and technology adoption are very closely linked (Barrett, 2008). Technologies help farmers to enter the market by allowing them to produce a marketable surplus, while the availability of market opportunities provides farmers with incentives to produce more or change their patterns of production, to add value to their production, and to innovate. Markets therefore strongly influence the technologies and practices adopted by farmers.

The linkages between market participation and innovation are becoming more important as income growth and economic liberalization change the conditions in which small family farms operate. A revolution in food supply chains has been under way in developing countries for more than three decades, involving extensive consolidation, very rapid institutional and organizational change, and modernization of the procurement system (Reardon and Timmer, 2012). Demand for high-value products, and the growing importance of integrating small farmers into value chains and trade can stimulate demand for small farmers' produce and provide incentives for innovation, while market failures and price volatility can be major disincentives to investment by family farmers. Regulatory policies concerning food safety and ecolabelling can also be drivers of innovation. The inclusion of small farmers in modern value chains could offer rural households market and employment opportunities. Governments should strive to establish the necessary regulatory instruments to bridge the significant gap in economic and political power that exists between family farmers and their organizations on one side, and the other contracting organizations on the other. The private actors and service providers involved in value chains often supply crucial inputs and services to family farms and represent an important source of innovation.

The inclusive business model approach, which includes poor people in value chains as producers, employees and consumers, represents a successful methodology for integrating farmers into modern value chains (Box 3). Other approaches include local food procurement from family farmers by different levels of government (local, regional and national).[33] Not only can public purchase schemes guarantee food security for vulnerable populations and income for family farmers, but they may also enhance collective action to strengthen family farmers' marketing capacities and ensure greater inclusiveness. Developing these market linkages requires investment in small and medium-sized food processors and small-scale traders at the retail and wholesale levels.

To enter commercial agriculture, farmers need not only to focus on technical innovation, but also to run their farms as businesses. This involves making management decisions on what to produce and where, and on how and to whom to sell. Farmers must also decide whether and how to compete in local or export markets, how to finance investments, how much to invest in product differentiation, how to organize farm production and how to join with their neighbours for collective action. Entering commercial agriculture therefore requires developing new kinds of individual and collective decision-making skills supported by advisory and business services.

For most smallholders, the transition from small-scale subsistence farming to innovative, commercial production is fraught with difficulties. Two types of barrier can hinder market entry by small family farms (Barrett, 2008). One is lack of access to productive assets, financing and technologies, which prevents farmers from generating marketable surpluses and adding value to their production; women farmers are particularly vulnerable to this barrier. Enabling small family farms to produce a marketable surplus, including through investment in productive assets and innovation, is a precondition for improved market integration of small family farms. The excessive transaction costs of engaging with markets, especially in remote areas, represent the second type of barrier that can often prove insuperable. Overcoming these barriers depends on making mainly public investments in physical and institutional market infrastructure. The development of effective producers' organizations and cooperatives is also important and can contribute decisively to reducing the transaction costs associated with market entry by generating economies of scale.

Arias *et al.* (2013) discuss the determinants of smallholders' participation in agricultural markets, focusing on the heterogeneity of smallholder producers, and outline how to formulate appropriate measures to facilitate improved market participation. They argue that attempts to improve smallholders' productivity will have limited success if smallholders' linkages to markets are not strengthened simultaneously, and that limited participation in markets is a result not necessarily of lack of commercial orientation, but of constrained choices in a risky environment. However, smallholders are heterogeneous and will react in diverse ways to new market opportunities. Key areas for integrating smallholders into markets include supporting inclusive market development, promoting farmers' organizations, enhancing market information and other support services, and helping smallholders to manage risk.

In summary, innovation in family farming is strongly linked to increased commercialization, with innovation and commercialization depending on and reinforcing each other. Efforts to promote innovation and enhance innovation capacity in family farming need to go hand in hand with efforts to improve market integration. However, it is important to recognize that not all family farms are alike and not all have the capacity for innovation in farming and for commercial production. Some family farms may find it more effective to pursue higher incomes and improved livelihoods through non-farm activities. However, the two options are not mutually exclusive, as some members of farming families may move into non-farm activities. Innovation linked to increased commercialization, and diversification of farm household incomes can take place in parallel and can be mutually reinforcing.

[33] For a description of the Brazilian experience see Graziano da Silva, Del Grossi and de Franca, 2010.

BOX 3
Inclusive business models

Inclusive business models "include the poor on the demand side as clients and customers, and on the supply side as employees, producers and business owners at various points in the value chain. They build bridges between business and the poor for mutual benefit" (UNDP, 2008). The term "inclusive business" was first coined by the World Business Council for Sustainable Development in 2005, and the concept has received growing interest (Tewes-Gradl *et al.*, 2013).

For companies, the inclusive business model approach can provide opportunities by developing new markets, driving innovation, expanding the labour pool and strengthening value chains; for the poor the approach can enable them to become more productive, increase their incomes, and generally empower them (UNDP, 2008). Clearly, the market conditions in which the poor operate can make such business models risky and expensive for companies. Major constraints include limited market information, ineffective regulatory environments, inadequate physical infrastructure, missing knowledge and skills on the part of the poor, and restricted access to financial products and services (UNDP, 2008). Businesses that create such models range widely and include large multinational companies, large domestic companies, cooperatives, small and medium-sized enterprises, and not-for-profit organizations (UNDP, 2010).

In agriculture, the inclusive business approach can promote smallholders' inclusion in value chains. According to the International Center for Tropical Agriculture (CIAT), "Linking smallholders with modern markets is not only a matter of strengthening farmers' skills and capacities to become better business partners. It also requires the private sector to adjust its business practices to smallholders' needs and conditions to stimulate sustainable trading relationships" (CIAT, 2012). FAO implemented this approach in 16 countries across Africa, the Caribbean and the Pacific, and showed that improved business relationships can strengthen farmers' access to inputs and financial and business services without overreliance on public and project subsidies. Working with a preferred buyer with the capacity to

In terms of their capacity for commercial production and innovation, family farms can be broadly classified as:

- large family farms, which are essentially large business ventures although they are managed by a family and use mostly family labour;
- small or medium-sized family farms that:
  - are already market-oriented and commercial, generating a surplus for the market (local, national or international); or
  - have the potential to become market-oriented and commercial given the right incentives and access to markets;
- subsistence or near-subsistence smallholders who produce essentially for their own consumption and have little or no potential to generate a surplus for the market.

These are very broad categories; the exact composition of farms and the relative importance of different farm types will vary from country to country. The categories may also change over time because of socio-economic mobility influenced by such factors as public policies and support, access to markets, and public and private investment. However, within these broad categories family farms will have differing potential for innovation and diverse needs for an agricultural innovation system (Box 4).

The large farms in the first category are the most effectively integrated into well-functioning innovation systems. Their most important needs are an enabling environment for innovation and production, adequate infrastructure, and public research in agriculture to ensure long-term production potential. They may forecast demand has also been effective in stimulating production. FAO is currently preparing a publication that will present the framework and rationale behind the inclusive business model approach, lessons from its application, and guidance on implementation in different market and commodity contexts.

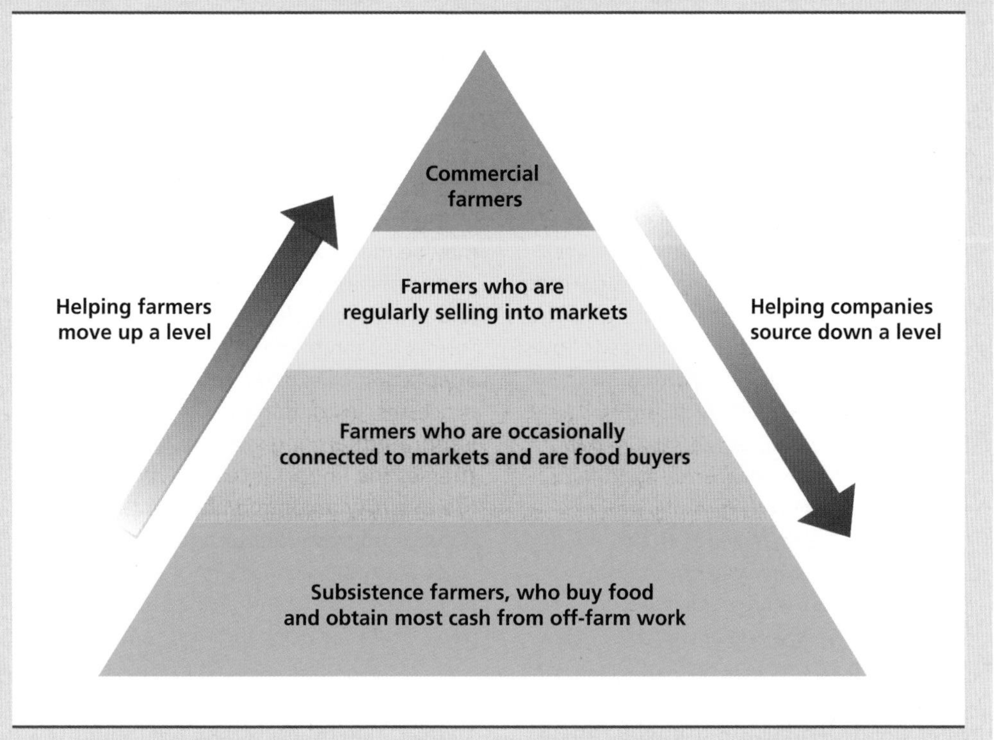

Source: Based on an original diagram prepared by Nicholas Sitko, Michigan State University, United States of America, for a presentation to the Agro-Enterprise Learning Alliance for Southern and Eastern Africa in 2010.

also need incentives to ensure that they apply sustainable practices and provide key environmental services.

Farmers in the middle category are less likely to be integrated into effective innovation systems, but have significant potential for innovation. In many countries, these farmers are likely to represent a large share of agriculture – in terms of land and number of farms. Promoting agricultural innovation in this group can have a major impact on food security and poverty alleviation and be transformative of world agriculture. Producers' organizations and cooperatives can play a central role in helping these farmers establish links to markets and value chains and integrating them into effective innovation systems.

Farmers in the third category have little or no capacity to produce a marketable surplus and are unlikely to be integrated into effective agricultural innovation systems. For these farmers, agricultural innovation can contribute to improved livelihoods and food security but, because their farms are so small and often remote, agriculture cannot be their sole or even main means of support if they are to live decent lives. Reaching millions of such very small farmers with relevant research, extension and innovation policies may be costly, hence the need to enhance social innovation and communication technologies to reduce costs. These farmers clearly need off-farm and non-agricultural livelihood options to supplement their farm incomes, and effective social protection to help them escape poverty. Overall rural development can enable them to diversify their sources of income and reduce their dependence on

BOX 4
What strategy should be taken towards small family farms?

Should governments support smallholder agriculture or larger farms? What are the best ways to improve food security and reduce poverty? Should strategies focus on smaller family farms? This is an old debate that continues today.

There is little agreement among development economists regarding the most effective government strategies for small farms. In a recent article, Larson *et al.* (2013) recognize a bias towards "institutional support for smallholder-led strategies", despite heated debate among agricultural economists regarding how appropriate such strategies are. The authors summarize the debate as follows:

> *...Collier (2008) charges that the development community has stressed less-innovative smallholder agriculture over more-productive commercial agriculture because of an overly romantic view of peasant farming. Hazell* et al. *(2010) counter that promoting smallholder agriculture is a more equitable approach to rural development, as well as a more efficient one. Lipton (2006) argues that emphasizing smallholder development partly compensates for policies in rich and poor countries that are, on balance, urban-biased.*

This edition of *The State of Food and Agriculture* recognizes the importance of sustainable productivity growth in small farm agriculture for poverty reduction and improved food security. It argues that there are two interrelated pathways along which small farmers' productivity may be increased: the development and application of new technologies and practices, including farmer-led and formal research; and the application and adaptation of existing technologies and processes, in combination with traditional integrated farming systems. It also stresses the importance of recognizing the diversity among family farms and the need to improve labour and other markets to provide supplementary or alternative forms of employment and income generation for poor farming families.

the income they generate from their small plots, and may also induce some of them to take up completely alternative employment opportunities.[34]

In conclusion, the diversity of family farms, both among and within countries, means that analysis and general policy recommendations are unlikely to be relevant for the entire category, whether they relate to innovation or other domains. There is need to differentiate and distinguish between different types of farm and different types of farming household within this broad category. It is also important to bear in mind that there are limitations to policies for encouraging innovation in agriculture. It may not be easy, cost-effective or even possible to reach all farmers in the family farm category. Alongside developing innovation capacity, there is a strong need to promote options for different livelihood strategies for farming families and their members, in the framework of broader rural development. Governments will need to develop their own strategies for different farm categories, based on their specific policy objectives, social and equity considerations and the costs of different options. For some governments, for instance, it may be important to support smallholder farming as a means of avoiding excessively rapid rural-urban migration; these governments may choose to focus support to innovation on very small farms. Others may wish to achieve similar objectives through policy instruments that focus on broader rural development.

[34] Fan *et al.* (2013) classify smallholder farms into three similar broad types: commercial smallholder farms, subsistence farmers with profit potential, and subsistence farms without profit potential. The authors argue that different strategies are needed for these different types of farms, depending also on the stage of development in the country. For subsistence farms without profit potential, the authors point to the need for education and training in non-farm employment as a key area of intervention.

Key messages

- Family farms are of critical importance to food security, poverty reduction and the environment, but they must innovate to survive and thrive.
- There are more than 500 million family farms in the world. They account for more than 90 percent of the world's farms and produce most of the world's food.
- These family farms are very diverse in terms of size, livelihood strategies and other characteristics, including their capacity to innovate in agriculture. This diversity means that innovation strategies must be designed to reflect the needs, constraints and capabilities of different types of family farm located in different socio-economic and institutional settings:
  - In low- and lower-middle-income countries, farms up to 5 hectares account for about 95 percent of all farms, occupy almost two-thirds of agricultural land, and produce the greater part of national food output. Even these small and medium-sized family farms are very diverse, as are the countries in which they are located.
  - In upper-middle-income countries, the size distribution of farms is highly skewed. A few large farms control vast tracts of land, while 70 percent of all farms are smaller than 5 hectares and together control less than 5 percent of the land. Innovation policies in such settings should carefully consider the role of farming in the livelihood and food security strategies of the smallest farms.
- Small and medium-sized family farms in low- and middle-income countries often have limited access to resources and low levels of labour productivity. At the same time, they also have major potential to increase their incomes and production through sustainable intensification.
- Access to markets is an essential driver of innovation in family farms. Improving the market integration of family farms that have the potential for commercial production is fundamental to promoting innovation.
- In addition to farming, most farming families – especially on small farms – depend heavily on non-farm sources of employment and income. Policies and programmes to promote innovation on family farms must go hand in hand with policies promoting overall rural development, to offer additional or alternative employment and income-generating opportunities in rural areas for farming families.

3. The challenge of sustainable productivity

Raising agricultural productivity in a sustainable way is indispensable for accelerating poverty reduction and feeding a growing world population from an increasingly constrained natural resource base. Farmers need to increase production on the available land to meet the growing demand for food. Many farmers also need to increase their labour productivity to make inroads into rural poverty. Farmers must also innovate to use natural resources more efficiently for environmentally sustainable production. This chapter reviews the challenge of sustainable productivity growth and assesses the opportunities and barriers facing family farmers in implementing more sustainable technologies and farming practices.

The need for sustainable productivity growth

Historically, agricultural productivity growth has allowed remarkable increases in food production, far outpacing growth in population and leading to a long-term downwards trend in real food prices. Over the last half century (1961–2011), global agricultural production more than tripled,[35] while the world's population expanded by 126 percent. Global cereal production grew by almost 200 percent, although the area harvested increased by only 8 percent. However, decreases in yield growth of major crops and recent rises in international food prices have led to renewed concerns over agriculture's ability to feed a growing world population, let alone to eradicate hunger (Figure 12).

It is still unclear whether the recent reversal of the downwards trend in prices represents a more permanent change. However, *The OECD-FAO Agricultural Outlook 2014–2023* (OECD and FAO, 2014) projects a short-term decline in international prices of agricultural products followed by stabilization at levels above those of the pre-2008 period. In a comparison of long-term scenarios for agriculture in ten global economic models by von Lampe *et al.* (2014), the different models show average annual increases in real global producer prices for agricultural prices ranging from -0.4 percent to +0.7 percent between 2005 and 2050. These figures compare with an average decline of agricultural prices of 4 percent per year between the 1960s and the 2000s. In all models, incorporating climate change effects leads to larger increases in prices over the same period (Nelson *et al.*, 2014).

Population growth and rising incomes in many developing countries will continue to fuel growing demand for agricultural products, especially high-value ones. Although the world's population is now growing more slowly, it is still projected to reach 9.6 billion in 2050, up from 7.2 billion today (United Nations, 2013). Most of the growth will be in developing countries, especially in Africa and South Asia, which have the highest incidences of undernourishment; population in the least-developed countries is expected to double to 1.8 billion. Increasing agricultural productivity and production in these areas of the world is imperative.

FAO has projected that to meet the increased food demand resulting from population and income growth, agricultural production will need to be 60 percent higher in 2050 than in 2006 (Alexandratos and Bruinsma, 2012). Pressure on increasingly scarce land and freshwater resources is expected to grow, as there is little scope for expanding agricultural land, except for in parts of Africa and South America. Much of the additional land that is theoretically available is either not suitable for agriculture or can be brought into production only at

[35] According to the FAOSTAT index of net agricultural production, which is net of intermediate production such as seed and feed.

FIGURE 12
Global food price index in nominal and real terms, 1960–2012

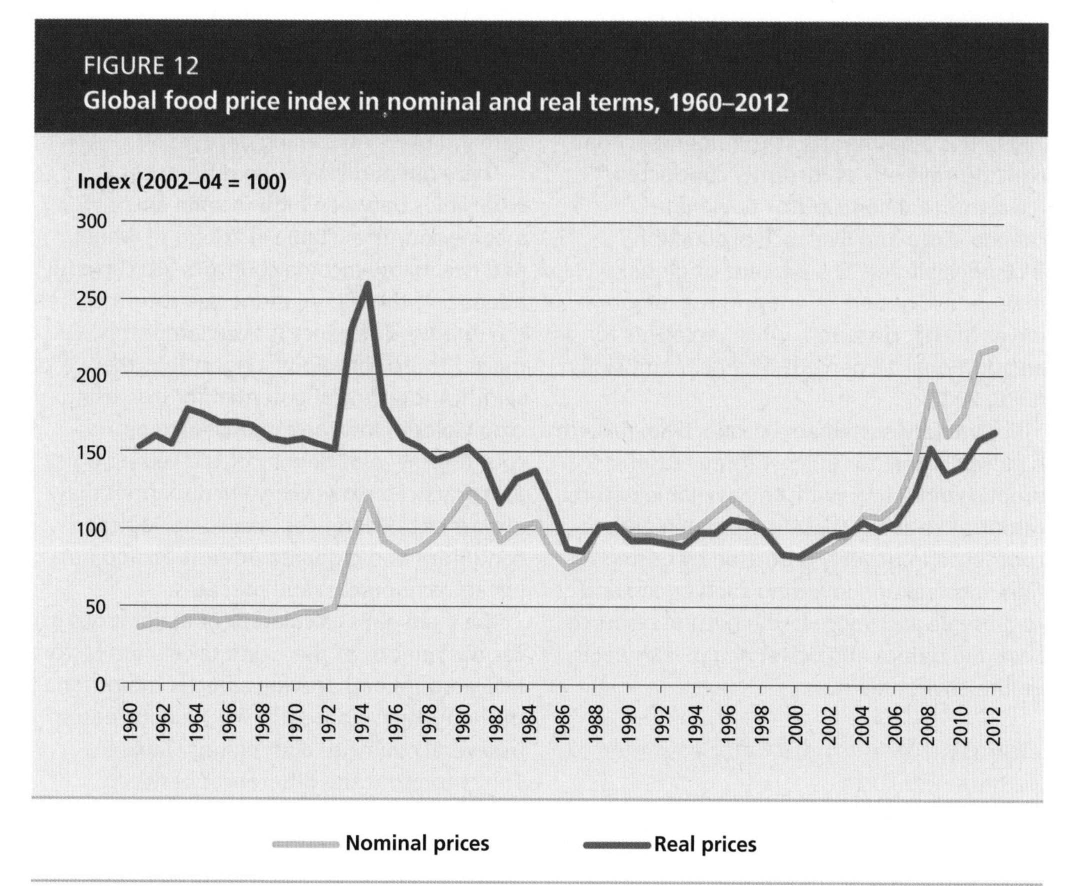

Notes: The World Bank's global food price index is calculated using prices for fats and oils, grains and various other food commodities. The index measures movements in international prices and not necessarily domestic prices. The World Bank's Manufactures Unit Value Index is used to deflate the nominal price index and produce the real price index.
Source: World Bank, 2013.

considerable ecological, social and economic cost. Most of the increased production must therefore be met through higher yields and increased cropping intensity (Alexandratos and Bruinsma, 2012).

In the past, agricultural production growth has often damaged land and water resources because of unsuitable management practices or deliberate choices to increase agricultural productivity at the expense of ecosystem services. Today, 25 percent of land is highly degraded and a further 8 percent moderately degraded (FAO, 2011a). Agriculture is by far the biggest user of water, and its current demands on the world's water resources are unsustainable. Inefficient use of water for crop production depletes aquifers, reduces river flows, degrades wildlife habitats and has led to salinization of irrigated land. By 2025, an estimated 1.8 billion people will be living in countries or regions with absolute water scarcity, and two-thirds of the world's population could be subject to water stress (Viala, 2008).

Biodiversity is also at great risk. The Millennium Ecosystem Assessment (2005) concluded that loss of biodiversity through human activities has been faster over the past 50 years than ever before in human history. Up to 75 percent of the genetic diversity of crops has already disappeared (Thomas *et al.*, 2004). Deforestation poses one of the gravest threats to biodiversity.

Climate change is another growing threat. Agriculture will suffer from the consequences of changing climate: rising temperatures, pest and disease pressure, water shortages, extreme weather events, loss of biodiversity, and other impacts. Negative effects on crop yields are more frequent than any positive impacts, and overall production is expected to continue to suffer, although there could be benefits in some places (IPPC, 2014). Production will also be increasingly variable. Developing countries – which are already more vulnerable to climate change because they are less equipped economically and technologically to defend themselves – will suffer more severe consequences than developed countries, and the gap between developed and developing countries will widen (IPPC, 2014; Padgham, 2009). It is also important to remember that agriculture itself, as currently conducted, is a significant contributor to climate change. Crop and livestock production is responsible for 13.5 percent of global greenhouse gas emissions and is a major driver of deforestation, which accounts for an additional 17 percent of global emissions (IPPC, 2007).

In summary, sustainable productivity growth is indispensable for at least three reasons: to produce more food with the available natural resources so as to meet growing demand; to contribute to poverty reduction by raising farm incomes and lowering food prices; and to preserve and improve the natural resource base and reduce and offset negative impacts on the environment.

Increasing land productivity to meet demand for food

While substantial additional amounts of food must be produced in coming decades without major expansion of cultivated area, growth in yields of major staple crops – wheat, rice and maize – at the global level has been much slower in recent decades than in the 1960s and 1970s (Figure 13). The question is whether yield growth rates can match the growth in demand over the coming decades.

There are also very large differences in crop yields between high-income and low-income countries (Table 4). Yields of wheat and rice in low-income countries are currently about half those in high-income countries; the relative difference is even larger for maize. These variations suggest that there is significant technical potential for increasing crop yields in low- and middle-income countries by adopting improved technologies and practices. However, yield disparities may also reflect differences in agro-ecological conditions and cropping intensities, and not just in technologies and practices.

The yield gaps calculated for major crops in various regions of the world take these factors into account and provide a better indication of the technical potential for yield increases in several countries and regions (Table 5). They represent the differences between current yields and those that could be obtained through optimization of inputs and management given existing agro-ecological conditions. Estimated yield gaps – expressed as a percentage of potential yields – exceed

FIGURE 13
Average annual rates of change in global crop yields, by decade and crop

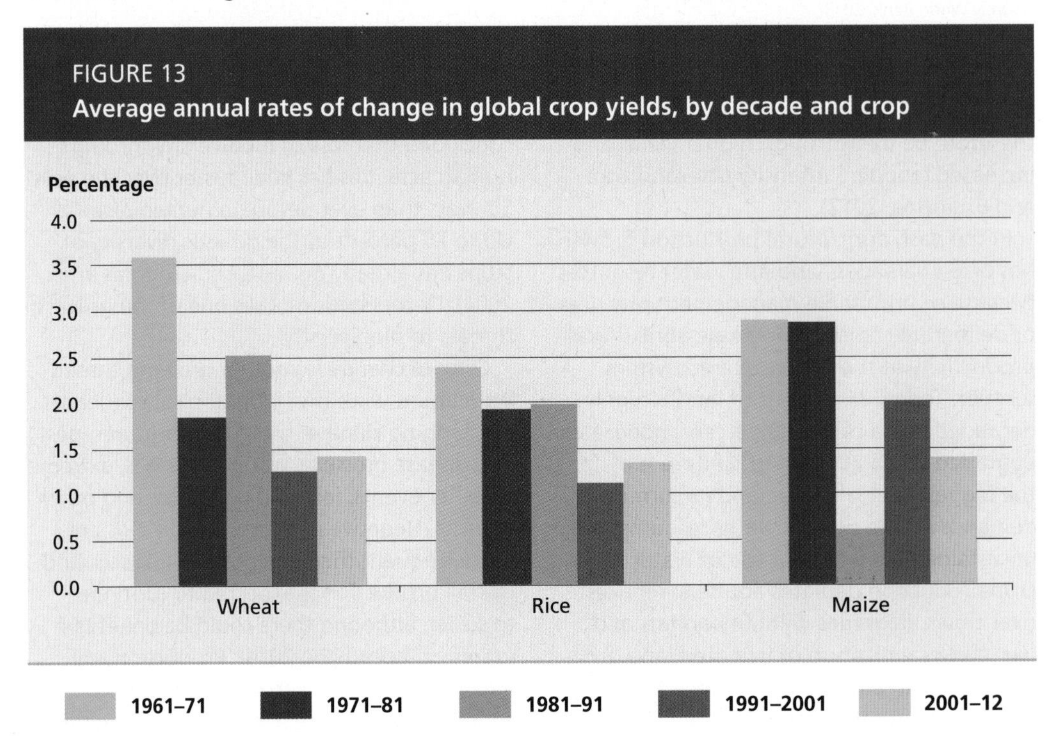

Notes: Rate of growth in crop yield (tonne/ha) is estimated using the OLS regression of the natural logarithm of crop yield on time and a constant term.
Source: Authors' calculations using FAO (2014b).

TABLE 4
Annual average crop yields, by income grouping, 2001–12

COUNTRY GROUPING	WHEAT	RICE	MAIZE
		(Tonnes/ha)	
Low-income countries	1.82	3.30	1.54
Lower-middle-income countries	2.74	3.65	2.74
Upper-middle-income countries	2.67	5.28	4.41
High-income countries	3.50	6.64	8.99
World	**2.92**	**4.16**	**4.87**

Notes: Country groupings are the same as those used by the World Bank (2012).
Source: Authors' calculations using FAO (2014b).

TABLE 5
Estimated yield gaps for major crops, by region, 2005

REGION	YIELD GAP
	(Percentage)
Sub-Saharan Africa	76
Central America and the Caribbean	65
Central Asia	64
Eastern Europe and Russian Federation	63
North Africa	60
Pacific Islands	57
South Asia	55
South America	52
Western Asia	49
Australia and New Zealand	40
Western and Central Europe	36
Northern America	33
Southeast Asia	32
East Asia	11

Notes: Crops included are: cereals, roots and tubers, pulses, sugar crops, oil crops and vegetables.
Source: FAO, 2011a.

50 percent in most developing region and are largest in sub-Saharan Africa, at 76 percent, and lowest in East Asia, at 11 percent. Reducing yield gaps could have high returns for food security, nutrition and incomes (Box 5). Reducing yield gaps for female farmers can have high returns as well (Box 6).

The higher prices on international agricultural markets experienced over recent years and projected for the future should provide an incentive for reducing yield gaps, both through increased use of inputs and the factors of production such as land and labour, and through the adoption of new technologies and practices. The capacity of family farms, especially small family farms, to respond to higher prices and increase their production depends on three factors: household access to assets, including natural resources, labour and capital; the degree to which the family farm is connected to markets; and the functionality of those markets, especially their integration with international markets (FAO, 2013e). Given their diversity

BOX 5
Impact of reducing yield gaps

OECD and FAO (2012) examined the possible effects of a hypothetical reduction of yield gaps by one-fifth between 2012 and 2021.[1] For cereals, the yield increases at the end of the projection period would amount to 7 percent for wheat and coarse grains and 12 percent for rice. Overall cereal production would increase by 5.1 percent. The increases in developing countries would be larger, while production would decline in developed countries. Another result of the increases in yields would be a 2.7 percent decrease in area harvested, as marginal land would be taken out of production.

The increased production would lead to major declines in world prices. For cereals, at the end of the projection period, prices would be almost 45 percent lower for rice and between 20 and 25 percent lower for wheat and coarse grains. Smaller but still significant declines would be recorded for oilseeds, vegetable oils and protein meals. The price reductions should be expected to have significant positive food security effects through improved access to food, even though 33 percent of the increased cereal harvest is projected to go into biofuel production. The effect on farm incomes could not be determined (as yields would increase while prices declined), but should vary across farm types and sizes. The authors nevertheless urge caution in interpreting the results, because the hypothetical yield increases are assumed to come at zero cost, i.e. solely through better management practices and improved seed varieties, but without increased fertilizer use.

[1] The impact was arrived at by comparing a baseline scenario for 2012–2021 in the Aglink-Cosimo model with a scenario in which crop yields increased relative to the baseline scenario in a manner that reduced the gaps proportionately by one-fifth in all developing countries by the end of the projection period 2012–2021. All the changes expressed are relative to the baseline values in 2021.

and heterogeneity, small family farms will be affected by these factors in different ways. Some smallholders are likely to intensify production on existing plots by adopting new technologies and practices, while others will increase the amount of land in production; however, some smallholders will be unable to benefit from improved opportunities because of their remoteness from and/or lack of participation in markets. Effective market linkages are essential for providing small family farms with the incentives they need to contribute to closing yield gaps.

Increasing labour productivity for poverty alleviation

As discussed in the previous chapter, reducing poverty in rural areas requires substantial increases in labour productivity – and thus rewards to labour input – on family farms. Globally, labour productivity in agriculture, measured as the total value of crop and livestock production per person employed in the sector, has been increasing over the past two decades, following earlier declines (Figure 14). Part of this growth may reflect an increase in physical output per worker and part a shift in production towards higher-value crops and livestock products.

However, labour productivity has been growing much more slowly in low-income than in high-income countries; as a result, the gap between high- and low-income countries is very large (Table 6). For the period 2001–2012, the value of agricultural production per worker in low-income countries was less than 3 percent of that in high-income countries (about 500 constant 2004–2006 international dollars per annum versus about 27 000). There is therefore great potential for labour productivity growth in low-income countries.

The widening gap in labour productivity between low- and high-income countries is largely because the rural labour force has been growing rapidly in low-income countries relative to opportunities for employment outside agriculture. Farmers in this country group have been using increasing amounts of labour on available land to increase output per hectare (Table 6).

FIGURE 14
Average annual rates of change in global agricultural labour productivity, by decade

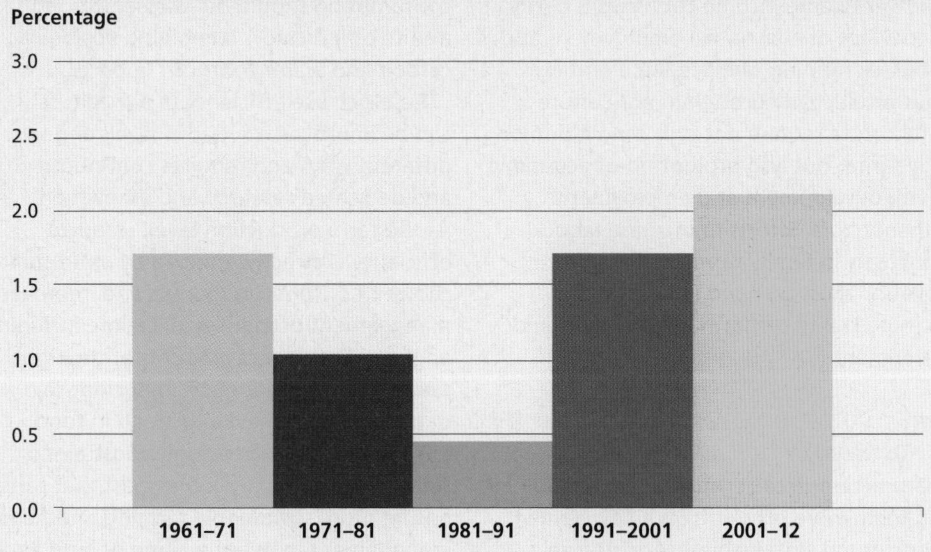

Notes: Labour productivity is the value of agricultural production per person employed in agriculture. Annual rates of change for the decade are estimated using the OLS method. The value of agricultural production is expressed in constant 2004–06 international dollars and is net of intermediate production such as seed and feed. For more details, see Notes on the annex tables.
Sources: Authors' calculations using FAO (2014b; 2008a). See Annex table A3.

TABLE 6
Average annual level and rate of change in labour productivity, by income grouping

COUNTRY GROUPING	AVERAGE LABOUR PRODUCTIVITY (2001–12)	AVERAGE ANNUAL CHANGE (1961–2012) IN:		
		Value of agricultural production	Agricultural workers	Labour productivity (value/worker)
	(Constant 2004–06 PPP dollars)		*(Percentage)*	
Low-income countries	490	2.5	2.0	0.4
Lower-middle-income countries	1 060	1.9	1.1	0.8
Upper-middle-income countries	1 450	3.8	1.3	2.5
High-income-countries	27 110	1.2	-2.6	3.9
World	**1 530**	**2.3**	**1.2**	**1.2**

Note: Country groupings are the same as those used by the World Bank (2012a).
Sources: Authors' calculations using FAO (2014b; 2008a). See Annex table A3.

As a consequence, land productivity has been growing much more rapidly in low-income than in high-income countries, but at the expense of slow growth in labour productivity. In high-income countries, production has grown much more slowly, but farmers have been leaving the sector rapidly and labour-saving technologies have been adopted, leading to significant growth in the productivity of the remaining farmers.

As increasing labour productivity in agriculture is crucial for poverty alleviation – because labour productivity is a key determinant of farm incomes – the widening gap between country groups underscores the importance of innovation to promote labour productivity growth. Innovation to boost incomes and reduce poverty is a high priority, particularly in low-income countries. Given the large number of small family farms in low-income countries, a focus on these farms is essential to achieve significant reductions in rural poverty.

Slow growth in labour productivity in low- and lower-middle-income countries is partly due to a lack of alternative employment and income for farming families. Accelerating labour productivity growth in agriculture will therefore require not only innovation on family farms, but also promotion of economic growth, development and employment in other sectors. Efforts to increase labour productivity in family farming through innovation must go hand in hand with policies to create off-farm employment and development.

Using natural resources more efficiently and sustainably

As natural resources become more constrained, using them more efficiently is a key element of agricultural sustainability. Agriculture uses many resources and affects the natural resource base in complex ways. Agriculture also often provides multiple outputs and services, which can include valuable ecosystem services. For example, in addition to providing protein-rich food, livestock in mixed farming systems often consume waste products from crop and food production, help control insects and weeds, produce manure for fertilizing, and provide draught power for ploughing and transport. An important function of ruminant livestock is converting biomass that is not digestible by humans, for instance from wastelands and semi-deserts.

Natural resource use efficiency refers to the amounts of natural resource inputs used to produce a given quantity of output. It includes both the quantity of resources used (e.g. hectares of land or litres of water) and the possible deterioration in the quality of natural resource stocks (e.g. soil erosion, biodiversity loss, nutrient runoff) (Place and Meybeck, 2013). Given the complexity of agricultural production and resource use, measuring resource use efficiency through a single metric is not appropriate; different metrics are likely to be relevant when considering different resources and outputs in different contexts. The level of greenhouse gas emissions per unit of food produced is an indicator that stimulates increasing global concern. In water-scarce areas, water use (amount and quality) per unit of product is a critical indicator. Galli *et al.* (2012) suggest that no single indicator can comprehensively monitor human impact on the environment, and argue that the environmental impact of production and consumption should be assessed through a suite of indicators combining ecological, carbon and water footprint impacts.

Resource use efficiency in agriculture can be improved at various levels and in different ways and requires continuous and dedicated research and innovation. At the farm production level, resource efficiency is directly affected by appropriate choice of outputs and inputs and improved management of input application, including applying the correct amounts at the right times. In crop production, reducing yield gaps is key to achieving growth in food output from an increasingly constrained resource base. Technologies exist that can ensure more sustainable farming and forestry management, prevent erosion of land and/or avoid pollution of water. However much more innovation is needed, with sharing of knowledge to allow adaptation to specific local conditions (United Nations, 2011); appropriate practices are generally very context-specific and knowledge-intensive. Close interaction among researchers, extension systems and farmers should thus be promoted to foster exchanges between science and traditional knowledge and experience (Place and Meybeck, 2013).

Family farming and sustainable productivity growth

Family farms are central to sustainable productivity growth in agriculture. As seen in the previous chapter, in many countries, especially low- and lower-middle-income countries, small and medium-sized family farms occupy a large share of agricultural land and are responsible for much national food production. They are therefore indispensable in both narrowing productivity gaps and ensuring sustainability of production. However, helping family farms to produce more, to increase their incomes and to do so sustainably represents a major challenge (Box 7).

Neither the old paradigm of input-intensive farming nor reliance on traditional practices alone can solve future problems

BOX 6

Closing the gender gap in agricultural productivity

Improving women's productivity can make a substantial contribution to raising overall agricultural production. Women comprise an average of 43 percent of the agricultural labour force in developing countries, ranging from 20 percent or less in Latin America to 50 percent or more in parts of Asia and Africa. Women's roles and responsibilities in agriculture vary widely according to regional social and cultural norms. However, one generalization seems to be valid everywhere: women farmers achieve lower yields than men farmers – not because they are bad farmers but because they have less access to everything they need to be more productive.

The State of Food and Agriculture 2010–11: Women in agriculture – closing the gender gap for development identified 27 studies that allowed direct comparison of yields between men's plots and women's plots. These studies covered a wide range of countries, crops, time periods and farming systems. The estimated yield differences ranged widely, but many clustered around 20–30 percent, with an average of 25 percent. The studies also found that the yield differences were fully explained by women's lower use of productive resources, such as improved seed varieties, chemical fertilizers, irrigation and other inputs (see, for example, Udry *et al.*, 1995; Akresh, 2008; Adeleke *et al.*, 2008; Thapa, 2008).

The vast majority of the literature confirms that women are just as efficient as men and would achieve the same yields if they had equal access to productive resources. However, almost universally, women have more restricted access than men to productive resources and opportunities – land, livestock, inputs, education, extension and financial services. Data from 14 nationally representative household surveys from all regions confirm this pattern of lower access (FAO, 2011b).

In addition, women and girls in rural areas bear a tremendous time burden for activities such as collecting fuelwood and water, which are essential to household well-being but prevent women from carrying out potentially more rewarding and productive activities. For example, women in rural Kenya, Uganda and the United Republic of Tanzania collect water an average of four times per day, spending about 25 minutes for each trip (Thompson *et al.*, 2001); and women in rural Senegal walk several kilometres a day carrying loads of fuelwood that weigh more than 20 kg (FAO, 2006).

Many of these tasks could be made much less onerous and time-consuming through the adoption of simple technologies. For example, the construction and rehabilitation of water sources in six rural villages of Morocco reduced the time that women and young girls spend fetching water by 50–90 percent, and was credited with increasing girls' primary school attendance by 20 percent over four years (World Bank, 2013). Similarly, the introduction of locally produced fuel-efficient stoves in western Kenya saved women about ten hours of work per month, with additional benefits in terms of improved indoor air quality and job opportunities in the production of stoves (Okello, 2005). Appropriate farm tools and improved seeds for women can also reduce the drudgery and time spent in the field, while helping to close the gender gap in yields (Singh, Puna Ji Gite and Agarwal, 2006; Quisumbing and Pandolfelli, 2010).

Closing the gender gap in access to productive resources could give an important boost to agricultural productivity and output and generate significant social gains. *The State of Food and Agriculture 2010–11* estimated that total agricultural output in developing countries could increase by 2.5–4 percent with significant benefits for food security.

Source: FAO, 2011b.

BOX 7
Sources of productivity growth

Agricultural output growth can be achieved in various ways. The two most common methods have been to use more inputs – including labour – per hectare, and to expand into new lands. However, both have often been associated with high rates of environmental degradation and low economic efficiency. The key to sustainable agricultural growth lies in growth in total factor productivity (TFP). TFP indicates that land, labour and inputs overall are being used more efficiently as a result of technological progress, adoption of innovative practices and human capital development.

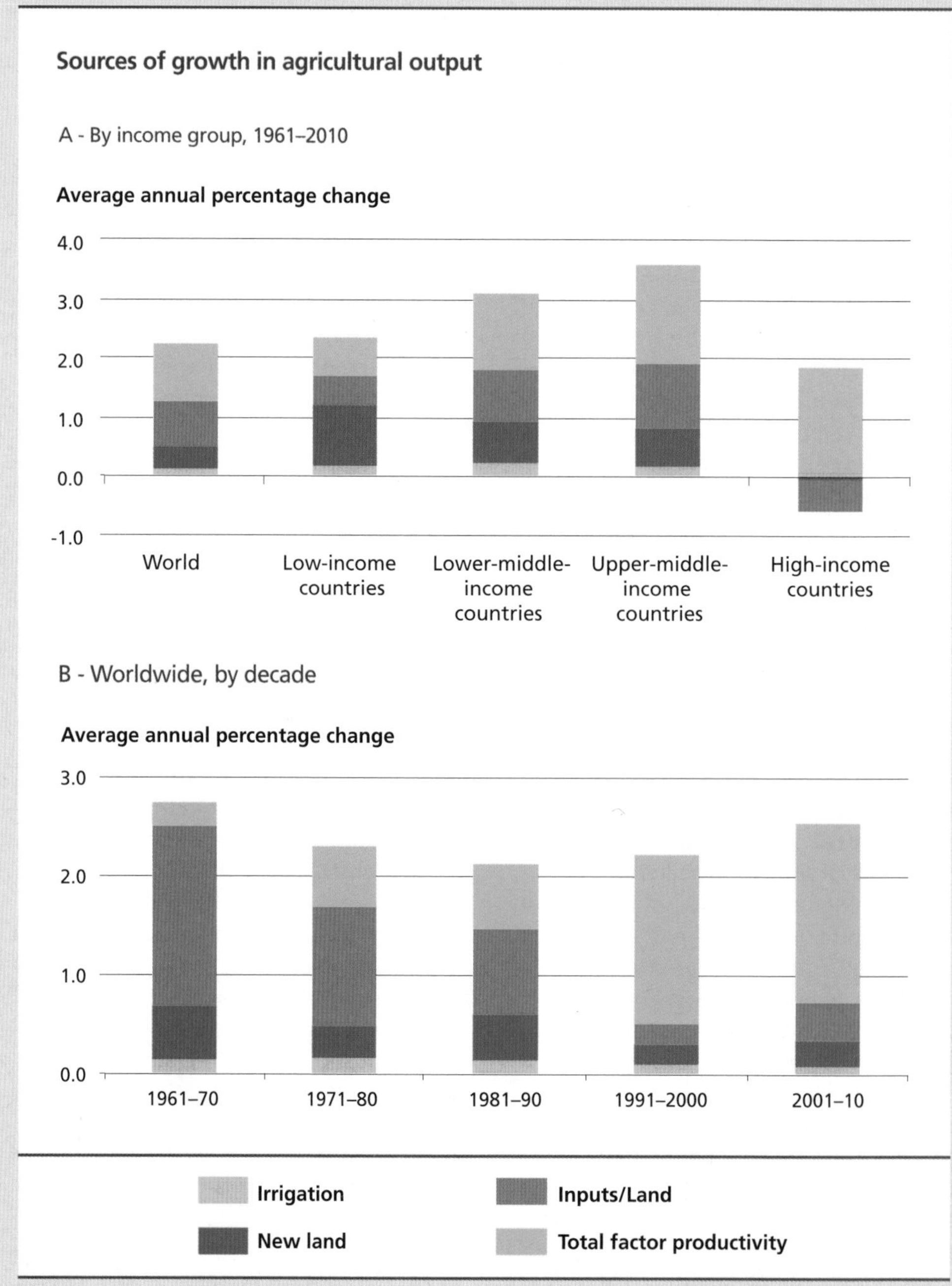

Sources: Calculations by Fuglie, using Economic Research Service (2013) and updated information presented in Fuglie (2012).

Fuglie (2012) decomposes sources of agricultural output growth over the last half century (Figures A and B) into four components: increased input use, including labour, per land area; expansion of irrigation; extension into new land area; and TFP. Globally, over the period 1961–2010, TFP growth accounted for about 40 percent of total growth in agricultural production (Figure A), becoming increasingly dominant over time (Figure B). In high-income countries, TFP growth has been the main contributor to agricultural output growth. In low-income countries, TFP growth has been modest, and most output growth has been achieved by expansion of agricultural areas. However, over the last decade TFP growth has increased significantly in low-income countries too.

In the long term, agricultural development must be based on sustained levels of TFP growth, which in turn depend on innovation capacity. Low levels of TFP growth in several developing countries, including in sub-Saharan Africa, present a clear challenge. In countries with a large proportion of small family farms, promoting innovation among these farms holds the key to ensuring TFP growth.

However, TFP growth does not by itself ensure environmental sustainability, and TFP estimates do not normally take into account the possible negative effects of agricultural activities on environmental resources. Environmental impacts such as biodiversity loss, nutrient runoff into water bodies, greenhouse gas emissions and other negative effects are generally not included in calculations of TFP (IFPRI, 2012), but they must be considered.

of sustainable productivity growth in the face of climate change. Future productivity growth in agriculture must be based on sustainable intensification (Box 8). Sustainable agricultural intensification practices are techniques that produce more output from the same area of land while reducing negative environmental impacts and enhancing natural capital and the flow of environmental services (Pretty, 2008; Pretty, Toulmin and William, 2011). Many such practices fall into the category of sustainable land management, such as soil conservation, improved water management, diversified agricultural systems and agroforestry. More conventional yield-enhancing technologies such as improved seed varieties and mineral fertilizers are also valuable options, especially if combined with greater attention to efficient use of these inputs.

Sustainable technologies and practices that have already been adopted and have generated large productivity gains in developing countries include low-tillage farming, crop rotation and interplanting, water harvesting and recycling, water-efficient cropping, agroforestry, and integrated pest management (United Nations, 2011). Other technologies hold promise for improving the resistance of crops to pests and extreme weather, reducing food contamination and reducing greenhouse gas emissions. However, farmers may need to be encouraged to apply such practices.

Family farms are generally part of larger productive landscapes, which frequently include forests, pastures or fisheries. Food security, nutrition, biological and genetic diversity, water and soil retention and recharge, pollination and a range of income-generating possibilities depend on these broader dimensions, and innovation must take them into account. Family farmers' decisions about their crop, livestock, fishery or off-farm activities, and the types of practice they use depend on their particular agro-ecological and market conditions, the incentives they face, and specific household characteristics such as wealth, education, age and gender.

To secure their livelihoods, households routinely make decisions on the allocation of productive resources to economic activities based on the relative return or benefit that each economic activity provides. The rate of transformation between allocated resources and outcomes depends on a number of conditioning factors, as well as the technologies employed. For sustainable agricultural intensification it is necessary to consider not only the agricultural output, but also possible environmental co-products, such as soil erosion or protection, greenhouse gas emissions, etc. Sustainable productivity growth encompasses not only the transformation of resources to agricultural products but also the extent to which environmental benefits or costs are co-produced with the agricultural system.

Benefits, costs and trade-offs of innovation for sustainable farming

Private returns versus public benefits

A major issue in sustainable agricultural intensification is whether there are trade-offs between productivity growth and economic returns to farmers on the one hand, and environmental benefits and ecosystem services on the other. Such trade-offs are frequent under the institutions that currently govern agricultural systems, in which environmental goods are generally not valued. For instance, reducing livestock numbers, or managing

BOX 8

Save and grow: a new paradigm for sustainable intensification of smallholder crop production

In its publication, *Save and grow* (FAO, 2011), FAO proposed a new paradigm of intensive crop production that is both highly productive and environmentally sustainable. FAO recognized that over the past half century, agriculture based on the intensive use of inputs has increased global food production and average per capita food consumption. In the process, however, it has depleted the natural resources of many agro-ecosystems, jeopardizing future productivity, and added to the greenhouse gases responsible for climate change.

Save and grow addresses the crop production dimension of sustainable food management. In essence, it calls for "greening" of the Green Revolution through an ecosystem approach that draws on nature's contributions to crop growth, such as soil organic matter, water flow regulation, pollination, and biocontrol of insect pests and diseases. It offers a rich toolkit of relevant, adoptable and adaptable ecosystem-based practices that can help the world's 500 million farming families to achieve higher productivity, profitability and resource use efficiency, while enhancing natural capital.

This ecofriendly farming often combines traditional knowledge with modern technologies that are adapted to the needs of small-scale producers. It encourages the use of conservation agriculture, which boosts yields while restoring soil health; controls insect pests by protecting their natural enemies rather than by spraying crops with pesticides; reduces damage to water quality through judicious use of mineral fertilizer; and uses precision irrigation to deliver the right amount of water when and where needed. The save and grow approach also builds resilience to climate change and reduces greenhouse gas emissions through, for example, increased sequestration of carbon in soil.

However, the adoption of such an approach requires more than environmental virtue alone: farmers must see tangible advantages in terms of higher incomes, reduced costs and sustainable livelihoods, and must be compensated for the environmental benefits they generate. Policy-makers need to provide incentives, such as rewarding good management of agro-ecosystems and expanding the scale of publicly funded and managed research. Action is needed to establish and protect rights to resources, especially for the most vulnerable people. Developed countries can support sustainable intensification by providing assistance to the developing world. There are also huge opportunities for sharing experiences among developing countries through South-South cooperation.

Source: FAO, 2011c.

manure to reduce nitrogen runoff to water or emissions to the atmosphere could benefit the environment, but would probably increase costs or reduce returns to the farmer.

In the absence of mechanisms for compensating farmers for providing environmental services and public goods, or for penalizing them for any negative environmental impacts of their farming practices, farmers will base their decisions exclusively on the private costs and benefits that they derive from the adoption of specific technologies and practices. Incentives are needed if agricultural systems are to provide greater environmental benefits, as farmers are not generally rewarded for doing so. The available policy options for ensuring that environmental benefits are incorporated into farm management decisions include financial penalties and charges, regulatory approaches, removal of perverse incentives that may unintentionally encourage unsustainable practices, and payment for environmental services (FAO, 2007).

However, the trade-offs between private returns and public environmental benefits are not universal; sustainability and increased production may be compatible through the adoption of appropriate practices. Power (2010) argues that trade-offs between production and other ecosystem services (or disservices) must be evaluated in terms of spatial scale, temporal scale and reversibility and that better methods for evaluating ecosystem services may increase the potential for win-win solutions; however, appropriate management practices are critical to realizing the benefits of ecosystem services and reducing disservices from agriculture.

Assessments in developing countries have demonstrated that farm practices that conserve resources can improve the supply of environmental services and increase productivity (FAO, 2011c). A review of 286 agricultural development projects in 57 poor countries showed how 12.6 million farmers had improved crop productivity while increasing water use efficiency and carbon sequestration and reducing pesticide use; crop yields increased by an average of 79 percent (Pretty *et al.*, 2006). In another study, Pretty *et al.* (2011) analysed 40 programmes in 20 sub-Saharan African countries where sustainable intensification practices were introduced during the 1990s and 2000s. The authors found that across the 12.8 million ha in these projects, crop yields rose by an average factor of 2.15, but it took from three to ten years to achieve these gains.

The magnitude and breadth of climate change impacts on agricultural systems, and the contribution of agriculture to greenhouse gas emissions make consideration of climate change issues, as well as national development and food security objectives, particularly important when determining the best agricultural intensification strategies for a specific location. It is also important to consider adaptation to climate change as well as mitigation through reduced greenhouse gas emissions and increased carbon sequestration. FAO has developed an approach that specifically considers the trade-offs among multiple objectives, together with the need for institutions, policies and investments to support innovation and the adoption of relevant agricultural practices (Box 9). The approach does not recommend specific technical solutions but provides tools for assessing different technologies and practices in relation to climate change mitigation and adaptation as well as national development and food security objectives. It will allow countries to make more informed choices based on their national priorities.

Short-term costs versus long-term returns

The timing of the associated costs and benefits can also be critical for farmers' decisions and capacity to adopt sustainable practices. Frequently, introducing new land uses or management practices leads to a temporary decline in net farm income because of upfront costs. This decline can prove a major deterrent to adoption, even when the new practices would lead to significant returns to the farmer in the long run. The inability to bear short-run costs to obtain long-term benefits is often the reason why farmers do not adopt practices that offer higher returns (Dasgupta and Maler, 1995; McCarthy, Lipper and Branca, 2011).

Even where there are substantial, private, long-run returns to sustainable practices, different types of cost may constitute significant barriers to adoption by farmers (McCarthy, Lipper and Branca, 2011). Direct

BOX 9
Climate-smart agriculture for food security

Climate-smart agriculture (CSA), as defined and presented by FAO at the Hague Conference on Agriculture, Food Security and Climate Change in 2010, is an approach for assisting countries in managing agriculture for food security under the changing realities of global warming. CSA addresses three objectives: (i) sustainably increasing agricultural productivity to support equitable increases in incomes, food security and development; (ii) increasing adaptive capacity and resilience to shocks at multiple levels (from the farm to the national); and (iii) reducing greenhouse gas emissions and increasing carbon sinks where possible. The relative priority of each objective varies across locations, so an essential element of CSA is identifying the relative food security, adaptation and mitigation effects of agricultural intensification strategies in specific locations. Such identification is particularly important in developing countries, where agricultural growth is generally a top priority. Often, but not always, practices with strong adaptation and food security benefits can also lead to reduced emissions or increased sequestration. However, implementation of these synergistic practices may involve higher costs, particularly for upfront financing. Building capacity to tap into sources of funding for agricultural and climate-related investment is therefore an important part of CSA.

Clearly, CSA does not imply that every practice applied in every location should generate triple wins, which may not always be feasible; instead, it implies that all three objectives must be considered, to derive locally acceptable solutions based on local or national priorities. The CSA approach is being developed and tested on the ground with national and local partners and is designed to align with and support the United Nations Framework Convention on Climate Change (UNFCCC) process. Since the introduction of the CSA concept, there has been a growing movement at the international and national levels for its adoption and scale-up; a global alliance for CSA is under development, and a regional CSA alliance for Africa has been established. Concerns have also been raised about CSA, which is sometimes perceived as implying one type of technological solution or focusing on linking smallholder farmers to carbon markets. While these are misconceptions of the approach developed and advocated by FAO, the issue is complicated by use of the term "CSA" by a wide range of stakeholders applying various definitions.

CSA does not constitute a recommendation for any specific technological solutions to address climate change; rather, the approach provides tools for assessing which technologies will deliver the desired results in different locations. Analysis for CSA starts with the agricultural technologies and practices that countries have prioritized in their agricultural policy and planning. Information on recent and near-term projected climate change trends is used to assess the potential for food security and climate adaptation of different practices under *site-specific* climate change conditions, and the potential need for adjustments in technologies and practices. Examples of such adjustments include modifying planting times and changing to heat- and drought-resistant varieties; developing and adopting new cultivars; changing the farm's portfolio of crops and livestock; improving soil and water management practices, including through conservation farming; using climate forecasts to inform cropping decisions; expanding the use of irrigation; increasing regional farm diversity; and shifting to non-farm livelihood sources (Asfaw *et al.*, 2014; FAO, 2010a; Branca *et al.*, 2011). The mitigation benefits of these prioritized options for food security and adaptation can also be assessed and used in an overall investment plan for CSA that links to both agricultural and climate finance, such as the Global Environmental Fund and the Green Climate Fund.

costs are the most obvious, and include *investment costs*, which cover expenditure on equipment, machinery, and the materials and labour required to build on-farm structures; and *variable and maintenance* costs, which are recurrent expenses, such as for seeds, fertilizers or additional hired labour.

Indirect costs are less obvious but can be even more important. They are related to foregone opportunities, transactions and risk. *Opportunity costs* represent the foregone income associated with allocating resources to one activity at the expense of another. These costs can often be quite high in the initial phase of adoption of sustainable practices and can extend for some time after. For instance, in many cases, adoption of improved practices may lead to temporary declines in levels of production and a consequent loss of income, even though previous production levels are eventually reached and surpassed.

Transaction costs include the costs of obtaining information, bargaining and negotiation, and monitoring and enforcement. Costs associated with searching for and processing information on various techniques and practices can be a significant barrier to adoption. Improving information and advice to farmers through effective advisory services and networks (including effective use of information and communication technology [ICT]) is critical in reducing these costs.

Risk costs are generally associated with uncertainty regarding the magnitude and variability over time of the benefits that the farmer expects to realize from adopting different practices. Adopting any new technology may be perceived as a risky investment, as farmers need to learn new practices and typically do not have access to insurance. Insecure tenure rights can increase the risk associated with investing in new technologies and practices, especially if the benefits take time to materialize.

Gender barriers to the adoption of sustainable production

Women face particular constraints in their ability to innovate and their access to information, inputs and services. Studies have found that women are often much slower than men in adopting a wide range of technologies, mainly because of the problems they face in obtaining access to complementary inputs and services (Ragasa *et al.*, 2014), (Meinzen-Dick *et al.*, 2014). In addition, some of the technologies promoted for enhancing productivity, adding value and saving labour, energy or costs do not benefit women or respond to their needs. Women generally have lower levels of education, less access to inputs, credit and information, and smaller plots than their male counterparts (FAO, 2011b). They have less capacity to incur direct, opportunity or transaction costs to implement new practices. Women are more likely to choose activities with lower risks but also lower returns (FAO, 2011b). In many countries, outmigration by men seeking to diversify household income emphasizes the importance of enhancing women's access to information, resources and markets.

Sociocultural norms and traditions may impose additional barriers to women, including by restricting their mobility and ability to engage in trading. For example, women often lack the cash to pay transport fares or purchase vehicles, and there is additional concern regarding the safety of women travelling long distances alone. In some countries, restrictive cultural traditions also circumscribe women's use of transport facilities (Starkey, 2002; Ragasa *et al.*, 2014). All of these challenges hamper women's capacity to innovate.

Very few technology adoption programmes address the specific limitations faced by women in given contexts (Meinzen-Dick *et al.*, 2011). It is particularly important to consider the time burdens of women's domestic chores. Potential solutions involve greater participation by women farmers in the design of sustainable practices, and related training. Labour-saving technologies that reduce women's chores, increase their labour productivity and give them greater control over the outputs of and incomes from their work will have considerable impact on the well-being of women farmers (Doss and Morris, 2001; Ragasa *et al.*, 2014). The need for labour-saving technologies is even greater in households affected by HIV/AIDS, as women often bear the double burden of producing food and caring for the sick. In sectors and areas where women suffer disadvantages because of gender norms, extension and other interventions to support the adoption of sustainable agricultural practices should look for ways of overcoming gender discrimination.

Facilitating the adoption of sustainable technologies and practices

What are the factors that determine farmers' adoption of practices for sustainable productivity growth, and what should be done to stimulate innovative behaviour by family farms? A few answers to these questions are illustrated in selected case studies from Africa (Box 10).

An important lesson is that there is no single approach to adopting technologies and practices for sustainable productivity growth on small family farms. Local agro-ecological conditions and climate play a central role in the selection and successful adoption of innovative approaches to farming. Households' socio-economic characteristics are also important. Technologies and practices therefore need to be relevant and suitable to local conditions and the requirements of the farmers involved. Linking farmers to researchers can help ensure the development of relevant options. Information for farmers on appropriate practices and available options is also important. Effective advisory services, and networks for sharing information and experiences are needed so that farmers can make more informed choices.

Access to markets is a key driver of innovation. As discussed in the previous chapter, the prospect of marketing additional output provides a strong incentive for farmers to innovate. Trading infrastructure and institutional arrangements allowing farmers to sell their products are therefore critical.

Household assets largely determine the extent to which farmers adopt new practices and the specific practices that they adopt. Wealthier households are better able to finance the initial costs of practices with longer pay-off periods and to face the risks involved in new approaches. Lack of financing and insurance against risk are therefore particularly constraining for small family farms with limited assets. Effective social protection can help to increase farmers' capacity to confront the hazards involved in applying new, more productive and sustainable practices. Tenure security is also important in motivating farmers to invest in improved practices (De Soto, 2002), especially those with benefits that are likely to materialize only after considerable time.

For several types of sustainable practice, environmental co-benefits are extremely important. It is unlikely that such practices are widely adopted without mechanisms for compensating or encouraging farmers. For activities that generate local public goods, local collective action may be the appropriate solution.

Last, but not least, gender is a fundamental issue, partly because some of the factors that constrain the adoption of more sustainable and productive practices by men farmers restrict women's adoption even more. Women farmers also face specific gender barriers that further limit their capacity to innovate and become more productive.

Institutions, especially local ones, are fundamental in addressing most of these issues and creating the right conditions for small family farms to innovate and apply technologies and practices that allow them to increase their productivity in a sustainable way. The effective functioning of local institutions and their coordination with both the public and private sectors, without excluding vulnerable family farmers, will strongly influence the capacity of small family farms to adopt improved practices. Strengthened producers' organizations can play a particularly important role in this respect. The challenge is to create an agricultural innovation system that helps small family farms introduce innovative and sustainable agricultural practices.

The following chapters examine some of these issues. The next two chapters deal with research and extension respectively, and how to make them responsive to the needs of family farms. The subsequent chapter looks at broader ways of promoting innovation capacity among family farms, both at the individual and collective levels and through the creation of an enabling environment.

Key messages

- Agricultural productivity must increase to meet the growing demand for food and to raise rural incomes. However, the natural resources that agriculture

BOX 10
Determinants of farmers' adoption of technologies and practices: case studies from Africa

In an analysis of what determines farmers' adoption of two conservation farming (CF) practices (minimum/zero tillage and planting basins) in Zambia, Arslan *et al.* (2013) found that extension services and rainfall variability are the strongest determinants. High rainfall variability increases the likelihood of adopting CF practices. Having the possibility of marketing output is also relevant, as the more selling points there are in a village, the more likely households are to adopt. Constraints to adoption include the limited potential for growing cover crops during the dry season in Zambia. The experience of CF adoption in Zambia illustrates that farmers select practices that are suitable to their agro-environmental conditions and that can be expected to secure increased marketable output in the presence of an institutional setting and available infrastructure for trading. However, extension services remain key to ensuring adoption of CF practices.

In Malawi, Asfaw *et al.* (2014) reviewed barriers to adoption of four agricultural practices that address climate change and other objectives (maize-legume intercropping, soil and water conservation, tree planting, and use of organic fertilizer), and two practices for improving average yields (improved maize varieties and use of inorganic fertilizers). Long-term climate patterns were found to play a significant role in the adoption of farm management practices. The findings also indicate that farmers choose technologies based on the specific characteristics of their plots and the overall wealth level of their households. For example, farmers with larger plots adopted practices with longer pay-off periods (soil and water conservation, maize-legume intercropping, and tree planting) but used less mineral fertilizer, which provides a more immediate return. Tenure security also makes it more likely that farmers adopt longer-term investment strategies.

In Ethiopia, Cavatassi *et al.* (2010) found that risk factors, coupled with access to markets and social networks, drive farmers' decisions to adopt modern varieties (MVs). Farmers appear to use MVs mainly to mitigate moderate risks, while the farmers who are most vulnerable to extreme weather events are less likely to use them. MVs appear best suited to more favourable production areas with adequate supplies of complementary inputs, while landraces appear to perform better than MVs in the production of subsistence crops under marginal conditions and with limited use of complementary inputs. Developing varieties that are more adaptable to climate change and extreme weather events will therefore become increasingly important for food security as climate change progresses. Preserving the richness of diversity within crops and promoting access to a diverse range of crop varieties may also be significant factors in facilitating farmers' ability to manage their risk, and social networks will have an essential role in providing such access.

depends on – land, water, biodiversity and others – are increasingly constrained and degraded, making it imperative that countries also preserve and restore the natural resource base.

- Countries may face difficult trade-offs between the objectives of agricultural productivity growth and natural resource preservation. Input-intensive production cannot meet the challenge of sustainability, while traditional low-input systems cannot meet the challenge of productivity growth. Future productivity growth must be based on sustainable intensification that combines increased productivity with conservation and improvement of natural resources.
- Family farms are central to overcoming the challenge of sustainable productivity growth, but must innovate to become more productive and must make their production more sustainable.

- Farmers often face barriers that hamper their capacity to innovate, including high initial costs of new practices and limited access to inputs, information, markets and technologies suited to their needs. Such constraints are often much more severe for women farmers, who have less access to productive resources and face significant social hurdles to innovation. Closing this gender gap can lead to major increases in sustainable agricultural productivity growth.
- Governments, international organizations and non-governmental organizations (NGOs) must help farmers overcome barriers to innovation for sustainable intensification. Secure property and tenure rights, transparent marketing institutions and good infrastructure are key elements of promoting the wider adoption of improved practices by family farms.
- Incentives may be needed to encourage farmers to adopt farming practices that combine increased production with environmental benefits and services. Locally developed knowledge needs to be supplemented with research and development suited to local agro-ecological and socio-economic conditions to provide farmers with suitable options for sustainable productivity increases.
- Local institutions such as producers' organizations can play a crucial role in facilitating family farmers' access to markets, capital, information and financing and in helping them to adopt improved practices. Effective participation of women in such organizations can help close the gender gap in access to productive resources.

4. Agricultural research and development for family farms

Farmers experiment and innovate continuously and have done so for millennia. Their efforts led to the domestication of the many crops and livestock species used in the modern food system. Formal scientific research in agriculture is a relatively recent phenomenon and has been largely responsible for the enormous growth in agricultural yields since the mid-twentieth century. Local indigenous knowledge – often implicit in farmers' practices – and formal scientific research should both be involved in the overall innovation system needed to enable family farms to achieve sustainable productivity growth and adapt to changing environmental circumstances. Building closer cooperation between formal and informal parts of the research system can help ensure that agricultural research and development (R&D) supports innovation by small family farms.

This chapter reviews the main international patterns and trends in formal agricultural R&D and makes the case for strengthening research efforts around the world. It analyses the potential for incorporating international research into national research systems and discusses new partnerships that combine the relative strengths of national and international, public and private, and formal and informal research efforts. Particular attention is paid to ways of orienting research towards the needs of family farms.

The importance of public agricultural research and development

Agricultural R&D requires sustained public investment for three main reasons. First, the results of agricultural research are often public goods, meaning they generate benefits for society beyond the value to the developer. Private researchers, including farmers themselves, therefore tend to underinvest in agricultural research with public goods characteristics. Second, as in many other branches of science, the results of agricultural research are cumulative, with current research building on past results (Box 11). This accumulation of research over time contributes decisively to productivity growth in agriculture (Pardey and Beddow, 2013). Third, there is often a considerable time lag – often of decades – between the expenditure of research funds and the benefits that the research may produce. Time is required both for achieving scientific results and for testing, adapting and widely adopting new technologies and practices. For this reason, Pardey and Beintema (2001) refer to investments in formal agricultural R&D as "slow magic".

An extensive body of literature has systematically shown that there are very high rates of return to public investment in agricultural R&D. This suggests that major gains could be achieved through increased public investment in research (Hurley, Pardey and Rao, 2013; Mogues *et al.*, 2012; Rao, Hurley and Pardey, 2012). The private sector can play a major role in certain types of agricultural R&D, especially in research with less pronounced public goods characteristics; but only publicly funded research is likely to produce the results needed to sustain productivity growth in the long run, especially in many low- and middle-income countries where incentives for private research in agriculture are weaker.

Changing patterns in agricultural research and development

Public investments

In spite of the importance of public agricultural R&D, growth in public expenditure slowed over the period 1970–2000, but has picked up somewhat during the past decade, except in high-income

BOX 11
The cumulative impacts of agricultural R&D

Evenson and Gollin (2003) assessed the impact of high-yielding varieties of 11 crops developed by the international agricultural research system (through the Consultative Group on International Agricultural Research [CGIAR]) and adopted in developing countries between 1960 and 2000, during the Green Revolution period of rapid agricultural innovation. The study highlighted important features of the development and adoption of agricultural technologies, most notably the cumulative nature of the process. The development of varieties suitable for conditions in developing countries was most rapid for crops such as rice and wheat, where developers could draw on advanced research previously undertaken in developed countries. For crops with little or no substantive prior research, such as cassava and tropical beans, it took much longer to develop suitable varieties. Nonetheless, by 2000 improved varieties had been developed for all 11 crops, with more than 8 000 modern varieties released by more than 400 public breeding programmes in more than 100 countries.

According to Evenson and Gollin, in many regions of the world, the adoption rate was quite rapid for most crops. In sub-Saharan Africa, however, the initial rate and extent of adoption were much lower, possibly because the varieties initially introduced from Asia and Latin America were not suited to local conditions. With the subsequent development during the 1980s of varieties that were better adapted to Africa, the rates of adoption increased, underscoring the importance of location-sensitive breeding.

Evenson and Gollin also estimated the contribution of high-yielding varieties to yield growth, crop production and food security. They found a very significant contribution in Asia and Latin America, which was stronger in the period 1981–2000 than in the previous decade. In sub-Saharan Africa, the contribution was significantly smaller but increased over the 1981–2000 period. The authors concluded that without the development of high-yielding varieties, crop yields would have been 19.5 to 23.5 percent lower; crop production would have been 13.9 to 18.6 percent lower in developing countries, but 4.4 to 6.9 percent higher in developed ones; crop prices would have been 35 to 66 percent higher, which would have contributed to crop area expansion with concomitant environmental effects; and calorie intake would have been 13.3 to 14.4 percent lower, with the proportion of children malnourished 6.1 to 7.9 percent higher.

countries, where research spending is already quite high (Figure 15). Upper-middle-income countries have seen a particularly sharp acceleration in expenditure growth in the last decade, largely because of rapid expansion of the public agricultural R&D budget in China.

An increasing share of public agricultural R&D is being conducted in middle-income countries, especially upper-middle-income countries (Figure 16), while public R&D is growing less rapidly in high-income countries. In 2009, low- and middle-income countries accounted for more than half of global expenditures on public agricultural R&D, but most of this spending is concentrated in very few large countries (Figure 17). For example, China, India and Brazil account for 19, 7 and 5 percent of global expenditures respectively. Together, these three countries plus the high-income countries account for 79 percent of global public spending on agricultural R&D, while the share of low- and middle-income countries is just 21 percent. Low-income countries' expenditures on agricultural R&D is particularly low, amounting to only 2.1 percent of the total in 2009, even less than their 2.4 percent share in 1960. Spending on agricultural research staff is an important indicator of long-term commitment to public R&D (Box 12).

FIGURE 15
Average annual rates of growth in public expenditure on agricultural R&D, by decade and income group

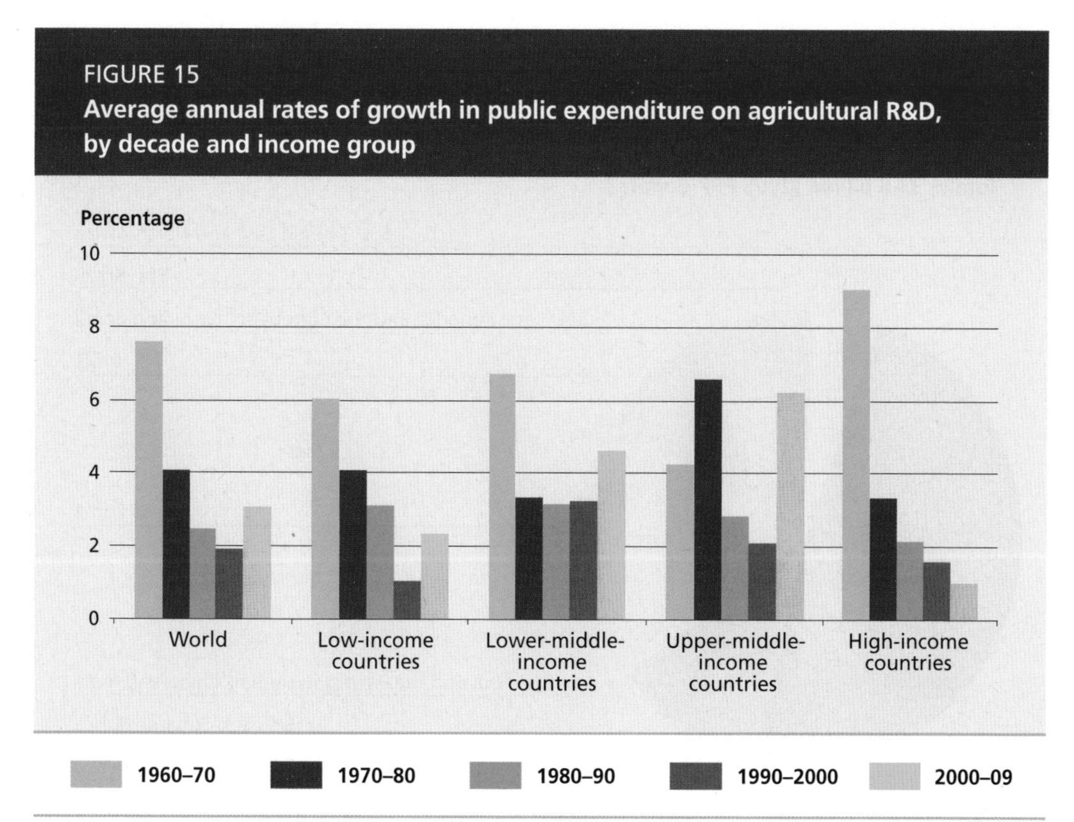

Notes: Simple average of annual rates of change in spending on agricultural research in each group for each decade. Data exclude countries in Eastern Europe and the former Soviet Union.
Source: Pardey, Chan-Kang and Dehmer, 2014.

FIGURE 16
Public expenditures on agricultural R&D, by income group

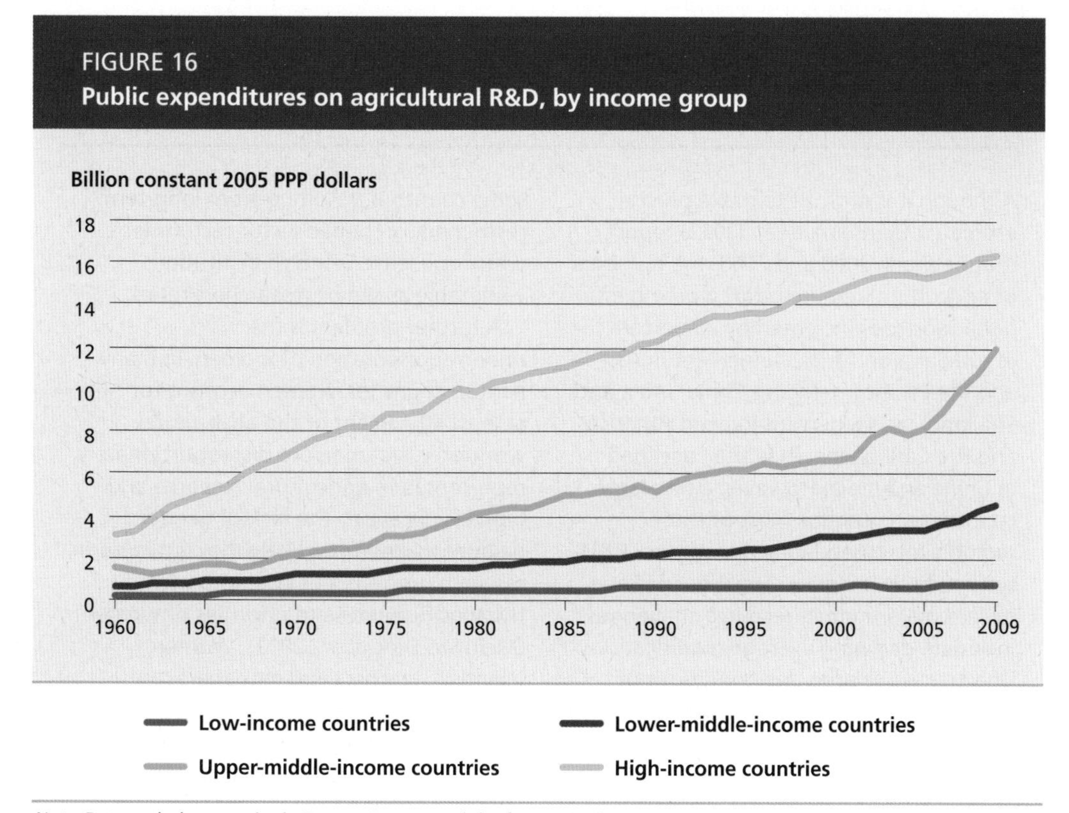

Note: Data exclude countries in Eastern Europe and the former Soviet Union.
Source: Pardey, Chan-Kang and Dehmer, 2014.

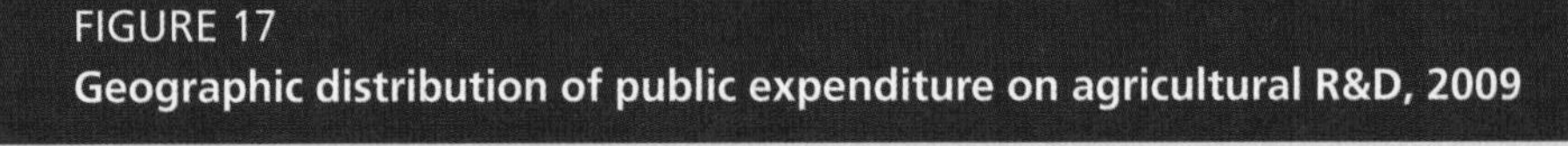

FIGURE 17
Geographic distribution of public expenditure on agricultural R&D, 2009

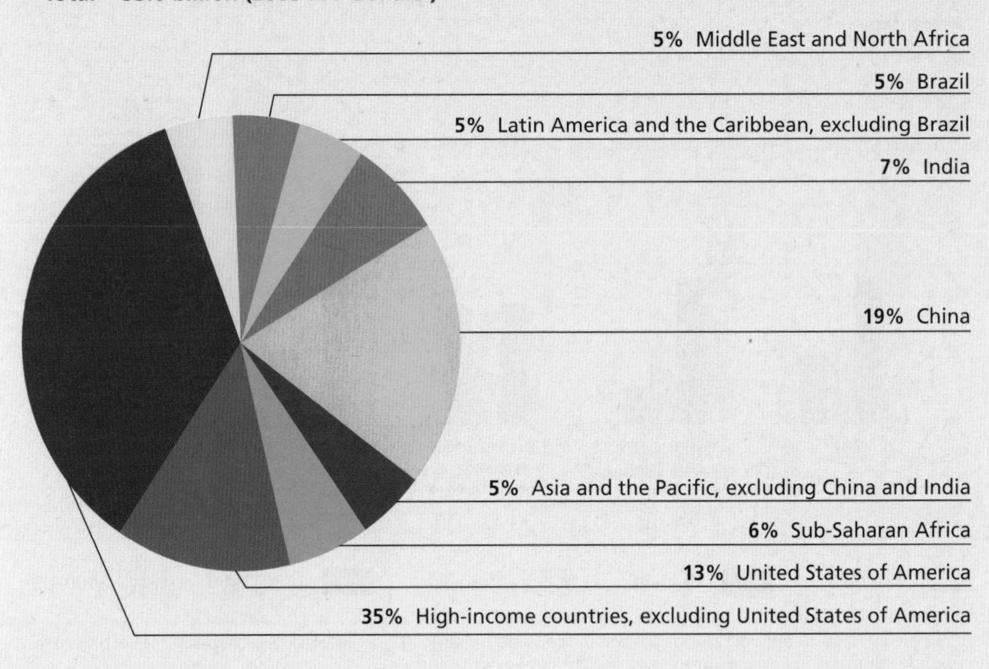

Note: All figures are rounded.
Source: Pardey, Chan-Kang and Dehmer, 2014.

BOX 12
Investing in agricultural researchers

Although it is difficult to make precise estimates, Beintema *et al.* (2012) report that between 2000 and 2008, the numbers of agricultural research staff working in public agencies increased by 25 percent in sub-Saharan Africa, 16 percent in Asia and the Pacific (excluding China, India and Thailand), and 5 percent in Latin America and the Caribbean, while they declined in China and India. However, a few large countries account for most of these regional increases. Many agricultural R&D systems in developing countries continue to face major human resource challenges, including declining average qualifications. Unattractive salaries and conditions of service make it difficult to recruit and retain qualified researchers, and many agencies have lost researchers to the private sector, CGIAR or richer countries. A particular issue is the rapidly ageing pool of scientists in some countries, resulting from long-term restrictions on public-sector recruitment, which will leave research institutions vulnerable as senior researchers retire.

A further problem is the underrepresentation of women. In many African countries, women account for at least 50 percent of the agricultural workforce, but men are disproportionately represented in agricultural research and higher education. The lack of gender balance makes it less likely that agricultural research programmes take into account the specific needs and priorities of women (Meinzen-Dick *et al.*, 2011). Women scientists, teachers and managers can provide different insights and perspectives from men, allowing research institutions to address the needs and challenges of both men and women farmers (Beintema and Di Marcantonio, 2009).

Private versus public investments

Private companies have long been involved in agricultural R&D. Although data are limited, private expenditure is estimated to account for 35–41 percent of total agricultural research expenditure (Pardey and Beddow, 2013); however, the vast majority of private research – perhaps 89–94 percent – takes place in high-income countries. Until recently, private agricultural R&D was concentrated in the mechanical and chemical sectors, where companies could develop proprietary products for the market; recent decades have seen increasing private investments in the life science sector, driven partly by changes to the governance of intellectual property rights for biological innovations, which make it easier for private companies to appropriate the returns on their investments (Wright and Pardey, 2006).

Beintema *et al.* (2012) (based on Fuglie *et al.*, 2011) estimate that private investment in R&D in agriculture and food processing increased from US$12.9 billion in 1994 to US$18.2 billion in 2008 (in 2005 purchasing power parity United States dollars). Primary agriculture accounts for less than half of this total, and its share has fallen from 51 to 46 percent. There is little information on private agricultural R&D in developing countries, but evidence from India (Pray and Nagarajan, 2012) and China (Pal, Rahija and Beintema, 2012) suggests that it has grown, and now accounts for 19 percent of total agricultural R&D spending in India and 16 percent in China (excluding food processing).

Although private-sector research is growing, there is still need for strong public-sector involvement. In developing countries, there are several disincentives to private agricultural R&D, including the high costs of serving small, remote farms, the difficulty of protecting intellectual property rights, unpredictable regulatory systems, and less developed value chains (Pardey, Alston and Ruttan, 2010). Much private research in agriculture builds on public research, which tends to concentrate on generating basic scientific findings rather than specific commercial applications (Pardey and Beddow, 2013). Public research is particularly important for generating science-based innovations in high-risk environments, and can also help maintain competitiveness in agricultural input markets that are characterized by increasing concentration (Fuglie *et al.*, 2011).

Investing in national research capacity

In many countries, public investments in agricultural R&D remain far too low relative to the sector's economic significance and importance for poverty alleviation. A commonly used indicator to assess countries' agricultural research efforts is the agricultural research intensity (ARI), which expresses national expenditure on public agricultural R&D as a share of agricultural GDP. Since the 1960s, ARI has increased substantially in upper-middle-income countries and very strongly in high-income countries (Figure 18), mostly because of the sector's relative decline in overall GDP. In low- and lower-middle-income countries, where agriculture accounts for much larger shares of income and employment, little progress has been made.

The higher ARI in high-income countries is partly because these countries have more knowledge-based economies and tend to emphasize basic and maintenance research to sustain high levels of productivity (Beintema *et al.*, 2012). In addition, public research agendas tend to broaden at higher income levels, where there is more emphasis on environmental and food-safety issues, while developing countries focus more on applied research to close productivity gaps and adapt technologies to local conditions (Beintema *et al.*, 2012).

There is no way to determine the "right" level of ARI. However, the United Nations Economic and Social Council's (ECOSOC's) resolution 2004/68, "Science and Technology for Development", recommends that governments increase their overall R&D expenditure for science and technology to at least 1 percent of national GDP. For the agriculture sector, countries in both the low- and the lower-middle-income groups are overall far from this target, although there are major differences within the groups. While some countries have well-managed and -funded systems, others – including some that are highly dependent on agriculture – have low and/or declining levels of R&D expenditures and capacity.

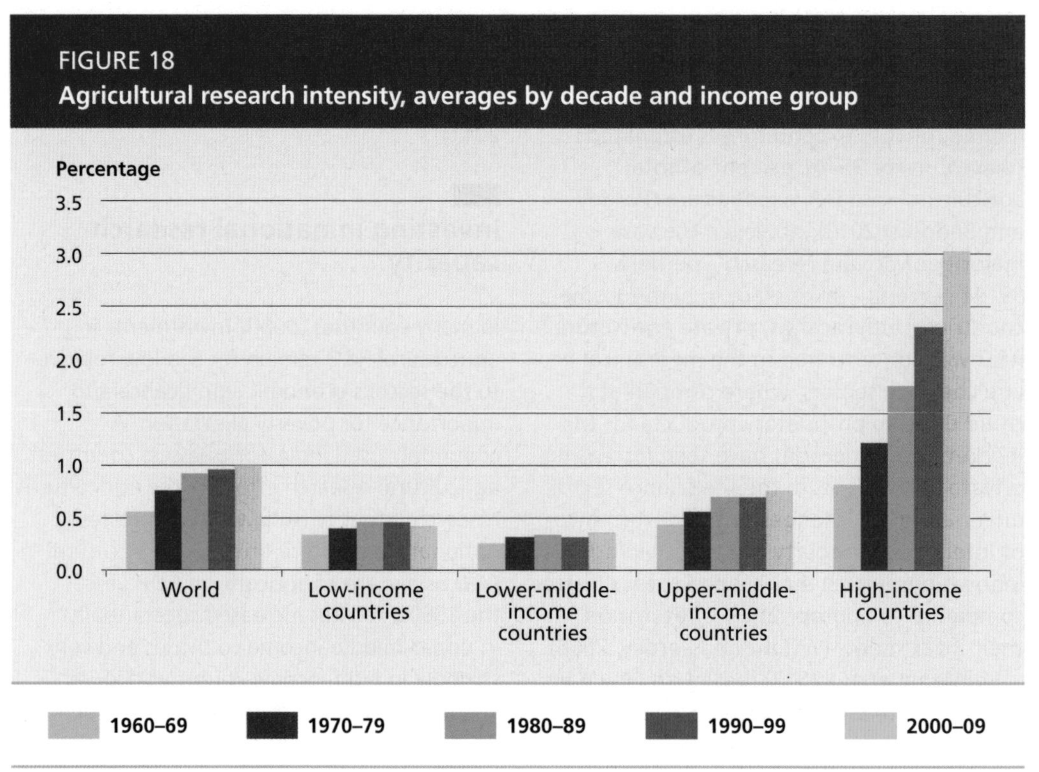

FIGURE 18
Agricultural research intensity, averages by decade and income group

Note: Simple average of annual agricultural research intensity.
Source: Pardey, Chan-Kang and Dehmer, 2014.

Funding public research and development

In many countries, the main mechanism for funding national agricultural research systems has traditionally been through block grants (also called core funding) from government. These grants are used to support research infrastructure, pay staff and enable strategic research programmes. In many countries, however, core funding today covers only salaries and not new investments for upgrading research facilities or for research costs. Discontent with traditional funding mechanisms and the perceived lack of effectiveness of agricultural research in general have led to the introduction of alternative funding methods.

For example, specific kinds of research can be directly commissioned from a provider. Through competitive grant schemes (CGS), funds can be allocated to innovative, high-quality and focused research proposals that are selected in a competitive and transparent manner (Echeverría and Beintema, 2009). This system has been used extensively in developed countries and, from the 1990s, some developing countries, such as in Latin America, where the World Bank has encouraged its use (World Bank, 2009).

Other new approaches include push and pull mechanisms. Push mechanisms reward potential innovations *ex ante,* while pull mechanisms reward successful innovation *ex post*. Models for pull mechanisms include prizes and challenge funds that reward achievements in technology development, such as high adoption rates, thereby creating strong incentives for researchers to select appropriate projects and focus on developing products that family farmers will want to use (FAO and OECD, 2012).

Nevertheless, stable institutional funding, including for infrastructure, is crucial for long-term research capacity (Box 13). Project-based funding can help to promote competition within the research system, but it has higher transaction costs. Newer research funding mechanisms such as CGS can be used to fund short-term projects, but should complement rather than replace institutional funding (Echeverría and Beintema, 2009). An evaluation of CGS and agricultural research in Brazil, Colombia, Nicaragua and Peru concluded that grants are most likely to make a sound and lasting contribution when they complement relatively strong public-sector involvement, and that to be able to compete, research institutions must have a minimum budget and a critical mass of staff (World Bank, 2009).

These new mechanisms for funding research can be important drivers in the innovation system. However, a major challenge for governments is to find a balance between funds for basic research and for applied research, and between stable, institutional funding and project- or programme-based funding tied to specific objectives and missions. Basic research requires a minimum number of qualified researchers, so small countries may prefer to prioritize applied research in allocating their limited national funds.

Partnerships for enhanced effectiveness of public research and development

As all countries have limited financial and human resources for agricultural research, they must allocate their resources strategically. Partnerships among national, regional and international research organizations can create synergies, as can better coordination and collaboration among researchers in the crop, livestock, forest, fisheries, natural resources and environmental sectors. National research institutes should also forge effective links with farmers, including smallholders and women, in order to respond better to local needs and conditions.

International partnerships

Basic scientific research findings can be transferred from one location to another and can be considered as global public goods while many findings from applied agricultural research must be adapted to local agro-ecological conditions and cultural preferences and constitute national or local public goods. Technology that has simply been transferred from other parts of the world or from international research centres, without local adaptive research, will have little value; all countries therefore need some degree of domestic research capacity (Herdt, 2012). Most countries rely on a combination of international and domestic research. The appropriate balance for a given country will depend on its stock of domestic research knowledge and its potential to take advantage of research results and technologies developed elsewhere ("spill-ins").

BOX 13
The importance of stability in funding agricultural R&D

Adequate levels of public funding for agricultural R&D are critical, but the stability of funding is also important. Stable long-term funding is essential for effective agricultural research, not least because of the time it takes for research projects to bear fruit. In the Agricultural Science and Technology Indicators (ASTI) global assessment of agricultural R&D, Beintema *et al.* (2012) estimated the volatility of R&D expenditures for 85 countries during the period 2000–2008. In low-income countries, average volatility was almost twice as high as it was in high-income countries, and considerably higher than in middle-income countries.

The highest volatility was found in sub-Saharan Africa, where many countries rely heavily on donors and development banks for their non-salary research expenditures (Stads, 2011). Funding from these sources is significantly more volatile than government funding. The completion of large donor-funded projects can frequently cause a financial crisis, forcing research institutes to cut back on programmes and lay off staff.

The ASTI study calls for a long-term commitment to agricultural research from national governments, donors and development banks. It calls on governments to identify their long-term, national R&D priorities and design relevant, focused and coherent programmes accordingly; recommends that governments diversify sources of funding and develop reserve funds or other mechanisms to avoid fluctuations in spending; and urges donors and development banks to align funding more closely with national priorities and to ensure complementarity and consistency among their programmes.

FIGURE 19
Agricultural spill-in potential *vis-à-vis* domestic knowledge stock

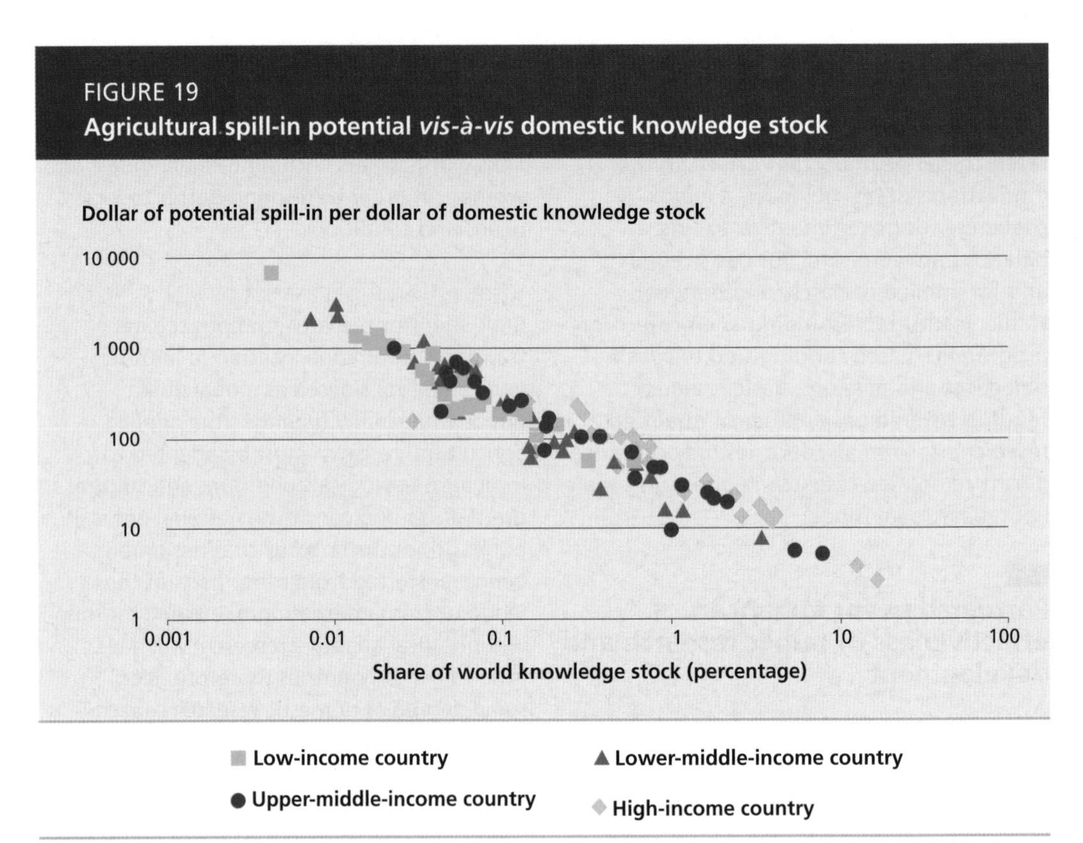

Notes: Excludes Eastern Europe and former Soviet Union countries. The horizontal and vertical axes are logarithmic scales in base 10.
Source: Calculations by Pardey using data from Pardey and Beddow (2013).

To guide such strategic choices, Pardey and Beddow (2013) have developed indicators of both the accumulated formal domestic knowledge developed by a country and the potential for spill-ins (Figure 19). Domestic stocks of productive knowledge arise from past research efforts. In Figure 19, the public stock of productive knowledge (calculated for 2009) represents accumulated R&D spending over the period 1960–2009, taking into account the delay between R&D spending and its impacts on productivity over time.[36] The potential for spill-in from other countries depends on the similarity in agro-ecological conditions and commodity mixes.

Figure 19 illustrates the relationship between home-grown knowledge stocks and the potential spill-in in a number of countries, by income group. Countries with high shares of the world's agricultural knowledge stock (the first axis) tend to have low potential for benefiting from agricultural knowledge from elsewhere – low spill-in potential (the second axis). These tend to be high-income or very large middle-income countries that can focus on domestic research and knowledge generation. In contrast, countries with a low share of the world's knowledge stock tend to have higher spill-in potential. These are mostly smaller countries and those with low per capita incomes. They would do well to focus their research efforts on adapting knowledge developed elsewhere for use by their own farmers.

The implication is that research from the rest of the world represents a substantial source of a country's potential to enhance productivity, particularly as a means of offsetting the historical underinvestment in agricultural R&D in lower-income countries. There is potential for increasing South-South cooperation in agricultural research between countries with larger public-sector research institutes – such as Brazil, China and India – and smaller national agricultural research institutes in countries with more limited research capacity facing similar agro-ecological challenges. It also underscores the importance of international research efforts that allow countries with limited domestic capacity to benefit from international research results and focus on adaptive research (Box 14).

[36] Applying results reported by Alston, Beddow and Pardey (2010).

BOX 14
International and regional investments in agricultural R&D

Most research at the international level is carried out by the CGIAR Consortium, which currently includes 15 centres. The first four of these centres were established in the late 1950s and 1960s with considerable financial support from the Rockefeller and Ford Foundations. During the 1970s, the number of centres grew to 12 and funding increased, resulting in a tenfold rise (in nominal terms) in total CGIAR investments. Funding continued to increase during the 1980s, but more slowly. During the 1990s, more centres were added but, although total funding continued to grow, average spending per centre declined. Since 2000, spending has again increased substantially, growing by 31 percent from 2000 to 2008 (in inflation-adjusted United States dollars) and a further 25 percent between 2008 and 2011 (Beintema *et al.*, 2012). In 2013, total CGIAR funding reached US$1 billion.

A number of other organizations and institutions engage in international research, mostly at the regional or subregional level. Since 2000, national agricultural R&D systems have established research networks such as the Association of Agricultural Research Institutions in the Near East and North Africa (AARINENA), the Asia-Pacific Association of Agricultural Research Institutions (APAARI), the Forum for Agricultural Research in Africa (FARA), the Forum for the Americas on Agricultural Research and Technology Development (FORAGRO) and the Central Asia and the Caucasus Association of Agricultural Research Institutions (CACAARI). These networks have enhanced collaboration and coordination of agricultural research activities and information sharing at the regional level. Some of them manage small competitive funding schemes (Beintema and Stads, 2011). The European Initiative for Agricultural Research for Development (EIARD) facilitates the coordination of European policy and support for agricultural research for development. Other recent initiatives are the World Bank-funded Eastern Africa Agricultural Productivity Project (EAAPP) and the West Africa Agricultural Productivity Program (WAAPP), which invest in regional approaches to agricultural research.

Numerous bilateral and multilateral initiatives now aim to develop agricultural innovation capacity in tropical countries. The Group of 20 (G20) recently launched the Tropical Agriculture Platform (FAO and OECD, 2012) to ensure better coherence and coordination among these initiatives, focusing on capacity development in the least-developed countries, over 90 percent of which are located at least in part in the tropics.

Public–private partnerships

In recent years, there has been growing interest in public–private partnerships (PPPs) involving governments, NGOs and the private sector. These novel institutional arrangements can be used to obtain access to additional financial and human resources, share risks and address other constraints in R&D (Box 15). The definition of PPPs varies throughout the literature, but they are generally considered to be collaborative relationships between public and private entities, with joint planning and implementation of activities to realize efficiencies, achieve joint objectives, and share benefits, costs and risks (Spielman, Hartwich and von Grebmer, 2007; Hartwich *et al.*, 2008).

However, public- and private-sector actors have divergent goals: public-sector organizations seek to maximize social benefits according to their mission statements; while private-sector actors aim to maximize profits (Rausser, Simon and Ameden, 2000). To ensure that both partners share the costs and benefits of conducting research, negotiations must focus on "defining goals, identifying complementary assets, and analysing the potential to segment markets for different partners" (Byerlee and Fischer, 2002). Overcoming cultural differences is one of the hidden costs of PPPs, which also include

BOX 15
A public–private partnership in biotechnology in Thailand

White leaf disease is a serious condition caused by phytoplasma – specialized bacteria that attack plants – in sugar cane. The disease is transmitted to the plant by the leafhopper *Matsumuratettix hiroglyphicus*. Weeds that grow in and around sugar cane farms are suspected carriers as they can be infected with phytoplasma and often show symptoms similar to sugar cane white leaf disease. To help combat this dangerous disease in Thailand's sugar cane industry, the National Center for Genetic Engineering and Biotechnology (BIOTEC) cooperated with the private-sector sugar producer and miller Mitr Phol Sugarcane Research Center – a subsidiary of the Mitr Phol Sugar Group – and an independent contractor to develop a rapid test for detecting white leaf phytoplasma in sugar cane. The detection method needed to be accurate, quick and simple to use, economical and non-perishable.

The project was divided into two phases. The first phase in 2005–2006 included R&D for an antibody able to detect white leaf disease. The second phase in 2007–2008 consisted in developing a white leaf disease test kit. Researchers from BIOTEC took the lead in the first phase, and the contractor carried out most of the design work in the second. BIOTEC provided all funding in the first phase and advanced 20 percent of project expenses for the second.

The white leaf disease test kits developed in the project proved to be innovative and valuable worldwide. They enable farmers to screen cane stalks for white leaf disease before planting. This not only reduces losses, but also minimizes spread of the disease to healthy plants. The kits have been commercialized domestically and internationally and sell for only THB500 (US$17) for a pack of ten, much less than alternatives. Mitr Phol and BIOTEC receive revenue and royalty fees from sales. Mitr Phol continues to promote use of the rapid test kit by sugar cane growers, with technical recommendations from BIOTEC regarding R&D in the sugar cane industry.

Source: FAO, 2013c.

the time costs of maintaining relationships, negotiating agreements and building trust among the partners (Spielman, Hartwich and von Grebmer, 2007; Rausser, Simon and Ameden, 2000). For the private sector, loss of control over intellectual property rights can be a significant concern.

PPPs often have extremely long lead-times between initial investments and the achievement of objectives. In the light of this and the relative novelty of PPP arrangements, there is as yet relatively little research documenting their effectiveness and impact.

Fostering research and development for family farms

Farmer-led innovation and formal R&D

Farmers are constantly experimenting, adapting and innovating to improve their farming systems. Indigenous knowledge is a major driver of "local innovation", which makes use of local resources, is site-appropriate and addresses the specific constraints, challenges and opportunities perceived at the local level (Wettasinha, Wongtschowski and Waters-Bayer, 2008). Local innovation engages local people in learning, inventing and adapting technologies and practices. Innovative farmers build on existing knowledge and share it with other members of the community. Understanding and supporting the processes of agricultural innovation and experimentation are important for enhancing sustainable productivity, which is strongly locality-specific (Röling and Engel, 1989; Long and Long, 1992; Scoones and Thompson, 1994)

Small-scale farmers and communities have shown great capacity to introduce productive innovations based on indigenous knowledge. These innovations have included developing seed varieties, designing soil and water conservation methods, and introducing post-harvest and value-adding technologies. Farmers have developed and used a range of land management practices to maintain and enhance soil fertility and productivity, including agroforestry, minimum tillage, terracing, contour planting, enriched fallow, green manuring, and ground cover maintenance (Critchley, Reij and Willcocks, 1994). Specific measures and technologies vary according to local biophysical, social and economic conditions.

However, scaling up and replicating these technologies is a challenge: farmer-led innovation is localized and confined to the bounds of farmers' knowledge and experience; indigenous knowledge is not uniformly spread throughout the community; and each individual possesses only part of the community's knowledge. Smallholder farmers very rarely document their knowledge, which is often implicit in their practices. Certain types of knowledge may be tied to economic or cultural roles within the community and may not be known by other community members. For example, studies in East Africa have shown that women usually possess remarkable knowledge about the qualities and uses of indigenous tree species and that many of those insights are unknown to men (Juma, 1987).

With changing circumstances – land pressure, new market opportunities, land deterioration – farmers' indigenous techniques may no longer be adequate. In situations where land is limited and the population continues to grow, traditional ways of farming may no longer be tenable. While most farmers practise some form of land management, changing biophysical conditions create the need for new technologies and measures for which farmers may lack the necessary knowledge base. Formal research can help to address this challenge by developing resistant cultivars; building knowledge about the life cycles of pests, biological control methods, suitable crops for erosion control and processes in nitrogen fixation; and designing more complex physical soil and water conservation measures.

Modern agricultural technologies and insights from research are crucial in providing farmers with guidance on addressing ecological concerns. For instance, science has a central role in mitigating or adapting to climate change. While plant breeders have been responding to climate-related stresses for a long time, climate change is making the development of new breeding activities and technologies even more important, to address challenges such as increased drought, higher temperatures, more widespread flooding, higher levels of salinity, and shifting patterns of pest and disease outbreaks.

In other words, local knowledge and traditional technologies are invaluable, but they cannot substitute for modern research and development: local knowledge and farmer-led innovation on the one hand, and formal research on the other must be seen as complementary. Understanding traditional agricultural practices and how they may be combined with new technologies and practices could lead to significant gains in productivity while mitigating the risks associated with change. Research for small family farms needs to take into consideration the close dependence on forests, fisheries, pasturelands and diversified livelihood systems of these farms. Combining scientific and traditional knowledge at the variety and landscape levels offers great potential.

Improving the linkages and cooperation between the formal research system and farmers can ensure that farmers' priorities are addressed, enhance farmers' access to and benefits from the work of researchers, and allow researchers to learn from and build on farmers' knowledge and innovations (FAO, 2012c). Producers' organizations can help facilitate these links. Researchers and extension workers should seek and encourage the involvement of farmers and their organizations in developing and adapting technologies to local farming conditions through interactive participation between professionals and farmers (Jiggins and de Zeeuw, 1992; Reijntjes, Haverkort and Waters-Bayer, 1992; Haverkort, Kamp and Waters-Bayer, 1991)

Research is being conducted in new ways to provide better support to innovation through collaboration (Thornton and Lipper, 2013). Many CGIAR centres have adopted new collaborative forms of germplasm development and diffusion involving different kinds of partners, such as the International Maize and Wheat Improvement Center's (CIMMYT's) MasAgro project, which is a partnership of more than 50 national and international organizations dedicated to improving sustainable agriculture. Other CGIAR centres, such as the International Centre for Agricultural Research in the Dry Areas (ICARDA), are using participatory approaches to crop improvement through variety selection in collaboration with national agricultural research organizations and NGOs. Recent partnerships with the private sector are leading to the uptake and diffusion of improved technologies that would not otherwise have been possible. In collaboration with national research organizations, some CGIAR centres are working directly with farmers' organizations and NGOs to select the most useful varieties and then bulking up supplies of quality seed and distributing it to farmers; for example, the International Crops Research Institute for the Semi-Arid Tropics (ICRISAT) is making small packets of seed commercially available to farmers.

Partnerships between researchers and family farmers

Traditionally, the role of extension systems was to link research to farmers through technology transfer. However, farmers have not always received technology that suited their particular environments and needs. New models of extension aim to ensure that there is two-way communication (see Chapter 5 for further discussion of new approaches in extension). Other approaches create closer partnerships between researchers and family farmers, such as Promoting Local Innovation (PROLINNOVA), which is an NGO-initiated multi-stakeholder programme, and other international projects such as the Platform for African-European Partnership on Agricultural Research for Development. Participatory approaches also offer important opportunities to ensure that women's needs and constraints are incorporated into technology development (Ragasa *et al.*, 2014).

Most participatory approaches for agricultural research have focused on adapting technologies to local conditions (Farrington and Martin, 1988). Numerous examples illustrate how involving farmers at different stages of adaptive research can complement the work of scientists (FAO, 2005). One example is participatory plant breeding (PPB), which has been incorporating farmers' active participation into plant breeding programmes since the 1980s. At least 80 participatory breeding programmes are documented worldwide, involving various institutions and crops (see FAO, 2009 for an overview). PPB allows farmers to select germplasm that is better suited to their environments, resulting in varieties that are well-adapted to the challenging lands typically worked by poor farmers (Box 16) (Humphries *et al.*, 2005).

PPB programmes may be formally led, with researchers obliged to complete research that is reproducible, or farmer-led, where farmers' needs for improved varieties drive the research programme, without any requirement for experiments to be replicable (Humphries *et al.*, 2005). Whether the programme is formally or farmer-led depends on the nature of the participation of both researchers and farmers. Participation can range from contractual, where one party maintains decision-making power and merely contracts the other for support, to consultative, collaborative or collegial, where both parties work together and share in decision-making (Vernooy *et al.*, 2009).

Evaluation of the impacts of PPB has been positive, showing that: i) PPB produces crop varieties that are more responsive to farmers' needs, thus increasing their adoption; ii) it does not appear to lower the cost-benefit ratios of breeding programmes; and iii) it accelerates the development of new varieties and their introduction into farmers' fields (Ashby, 2009). PPB programmes may also have other benefits in rural communities, such as strengthening social capital through farmers' associations and other networks, and providing educational opportunities for farmers (Humphries *et al.*, 2005).

Few impact assessments are broken down by gender: some studies highlight positive impacts on women and the benefits of involving women in PPB programmes, while others cast doubt on the gender impact of PPB (Ragasa *et al.*, 2014). Gender-sensitive targeting and programme design are needed, to support and facilitate the participation of women and to ease their specific problems with mobility, transport, time burdens and social constraints (Ragasa *et al.*, 2014)

BOX 16
Participatory plant breeding in Honduras

In Honduras, small farmers face high rates of rural poverty and inequality in land access. Wealthier individuals typically own the flatter, larger landholdings, leaving small farmers in remote areas to farm small plots on steep hillsides that are prone to erosion and poor soil fertility (Humphries *et al.*, 2005; Classen *et al.*, 2008). The concentration of infrastructure development in the north and centre of the country leaves many of these smallholders with few roads and markets and limited communication infrastructure. These factors, coupled with very traditional gender roles that discourage women from participating in agriculture, have restricted the development of social capital (Classen *et al.*, 2008). Typically, remote farmers have not been targeted by publicly funded research or extension, so many of them still use old techniques that worsen environmental problems; at high elevations, however, farmers' landraces outperform newer varieties (Humphries *et al.*, 2005). This combination of factors provides a unique opportunity for PPB programmes.

To improve the selection of varieties available to bean farmers in Yorito, Honduras, a PPB programme was implemented between 1999 and 2004. Participants included elected farmer research committees, known by their Spanish acronym as CIALs; the Foundation for Participatory Research with Honduran Farmers (FIPAH), a Honduran NGO that provides agronomic support to CIALs; and plant breeders from the Pan-American Agricultural School of Zamorano (Humphries *et al.*, 2005). Farmers were trained in experimental methods, and parallel trials were run at Zamorano. Early in the project, farmers were involved in selecting genetic materials that met their criteria for yield, disease resistance and commercial attributes. FIPAH agronomists served as facilitators and provided training to farmers in their communities.

In 2004, farmers selected a variety for release and called it Macuzalito, which is the highest point in the municipality of the four communities participating in the project. Farmers have since asked breeders to look for materials to cross with Macuzalito, indicating that they view PPB as a long-term commitment and process (Humphries *et al.*, 2005). Researchers at Zamorano who were once sceptical of PPB are now convinced that farmers are in the best position to choose varieties for their specific environmental and community conditions, and recognize that the skill sets acquired by CIAL members present opportunities to conduct research in areas that were previously inaccessible (Vernooy *et al.*, 2009). The PPB programme has increased the participation of women and built social and human capital in the communities; an assessment by Classen *et al.* (2008) indicates that CIAL members are more likely to join other associations and undertake continuing education.

Overall, the project has been successful in improving the livelihoods of the most marginal bean farmers on the Honduran hillsides. However, it should be noted that PPBs face several barriers. For example, a similar project in the Lake Yojoa region proved unsuccessful because the lake is much closer to a major urban centre. People found it easier to move between their farms and the city, making it difficult to ensure the stable membership required for a long-term PPB project.

Communication and collaboration between farmers and researchers involves a number of challenges. Farmers may not know what is expected of them in a research setting and may not be able to communicate clearly the tools, processes or products they require. The research system may not have the capacity to listen to and accommodate the multiple and diverse voices of family farmers. Scientists may find that their academic careers are advanced more readily through scientific publications and interactions with other scientists than by working in participatory research activities.

Research institutions may prioritize research avenues for which donor funding is available. Researchers and farmers alike may be unwilling to invest time, effort and money in talking to each other unless they see a clear advantage (FAO, 2012c).

Brokering or facilitation may therefore be needed to ensure that farmers and researchers cooperate. A recent example is the *Systèmes de production biologique diversifiés* (Syprobio – Diversified Organic Production Systems) project in West Africa, which required time and money to overcome such challenges through a patient cross-disciplinary approach (FAO, 2012c). Other examples of participatory research programmes are documented in FAO (2012d). One strategy for linking farmers to researchers is to increase the numbers of "transfer specialists" in research institutes, with some researchers from the institutes working more closely with extensionists, producers' groups and lead farmers to link research to local demands (Box 17).

Such facilitation mechanisms can help to develop partnerships between research and family farmers, but incentives are nevertheless crucial. These incentives could include policy and institutional changes that reward researchers for practical impacts in their research fields rather than for pure academic achievements, or that link the provision of research funding to teamwork with farmers (World Bank, 2012b).

Key messages

- Public agricultural R&D is particularly effective in promoting sustainable agricultural productivity growth and alleviating poverty. The benefits of public agricultural R&D are felt through three main channels: higher farm incomes, increased rural employment, and lower food prices for consumers. An extensive body of empirical evidence confirms the high returns to public investment in agricultural R&D.
- Private investment in agricultural R&D is growing rapidly, primarily in high-income countries but also in some lower-income countries. As private agricultural R&D focuses on products with a commercial market, public-sector investment remains indispensable to ensure adequate research investments in

BOX 17
Promoting technology transfer specialists in the Dominican Republic and Mexico

Two recently approved agriculture innovation programmes in Mexico and the Dominican Republic, supported by the Inter-American Development Bank (IDB), aim to strengthen the ties between research and extension through *Transferencista* (technology transfer specialists and researchers). Similar in role to the United States Land Grant research State Specialists, the *Transferencistas* are research professionals with a primary responsibility for ensuring that research is relevant to both extension professionals and farmers. The United States State Specialist model recognizes that different incentives, staff training, budgets and institutional mechanisms are needed for research that is useful to small farmers (Deller and Preissing, 2008).

In Mexico and the Dominican Republic, governments and IDB identified a lack of physical capacity, training, resources and incentives for research and extension to promote innovation. The two projects provide new resources to train and/or hire researchers as technology transfer specialists, upgrade training centres, train extension agents, develop mechanisms and tools to capture demand better, and develop metrics that better recognize the contributions of technology transfer specialists in the innovation agenda (Falconí and Preissing, personal communication, 2012). In Mexico, 32 outreach centres are being upgraded and staffed with technology transfer outreach specialists, and 90 researchers are being trained in participatory research methods. In the Dominican Republic, three outreach centres will be upgraded.

areas that are of little or no commercial interest to the private sector, such as the "orphan crops" that smallholder farmers grow in marginal areas of developing countries, or sustainable production practices.

- Countries must maintain, and in many cases increase, expenditure on agricultural R&D to ensure continued productivity growth and environmental sustainability, but the stability of public funding is also important for agricultural R&D to be effective. Innovative funding mechanisms can contribute, but stable institutional funding is also needed to ensure core long-term research capacity.
- Agricultural R&D can be strengthened through partnerships between national and international research agencies, between the private and public sectors, and among sectoral research institutes. Basic scientific research is needed to enhance the overall long-term potential for sustainable production but, because the results of such research are international public goods, international public research institutes may be better placed to carry it out. More adaptive research is needed to exploit this potential fully in the specific agro-ecological conditions in different countries. Countries with limited financial resources may thus choose to build on research results from larger countries or international institutes and focus their own efforts on adaptive research.
- There is potential for increasing South-South cooperation in agricultural research between countries with larger public-sector research institutes and smaller national agricultural research institutes in countries facing similar agro-ecological challenges.
- Farmer-led innovation and formal research are complementary; combining traditional knowledge with formal research can yield truly innovative approaches to support sustainable productivity growth among family farms. Farmers' participation in formal R&D projects helps ensure that the resulting technologies fit their real needs and builds on their experiences, but the professional incentives currently facing research organizations may not foster such collaboration. Producers' organizations and other forms of collective action can facilitate better communication and collaboration between farmers and researchers.
- Governments have a responsibility to help produce research that is relevant to the special needs of small family farms and to ensure proper governance of partnerships and collaborative efforts.

5. Agricultural extension and advisory services for family farms[37]

Agricultural extension and advisory services are central to achieving sustainable productivity growth among family farms. By facilitating farmers' access to information, such services can help reduce the gap between potential and actual yields and improve farmers' management skills (Anderson and Feder, 2007). They can help agriculture become an engine of pro-poor growth and equip small family farms to meet new challenges, including access to markets, adoption of environmentally sustainable production methods, and responses to climate change (Birner *et al.*, 2009). However, far too many family farms do not have regular access to extension services.

Although recent decades have seen the emergence of more pluralistic agricultural extension and advisory service systems, with private firms, producers' associations and civil society playing more active roles alongside traditional public-sector providers (Sulaiman and Hall, 2002), there is still an important role for government. In common with agricultural research, agricultural advisory services generate benefits for society that are greater than the value captured by individual farmers and commercial service providers, such as increased productivity, improved sustainability, lower food prices and poverty reduction. These public goods call for the involvement of the public sector, for example in providing advisory services to small farms and services to support sustainable production practices. The public sector also has a responsibility to ensure that the advisory services provided by the private sector and civil society are technically sound and socially and economically appropriate. This chapter discusses trends and challenges in agricultural extension and advisory services and their implications for small family farms.

Trends and patterns in extension

Studies have shown that investments in extension – in common with investments in agricultural research and development – have delivered high rates of return. In a review of extension programmes, Evenson (2001) found that although rates of return to extension varied widely, they exceeded 20 percent in three-quarters of the 81 extension programmes considered. In a survey of quantitative studies of rates of return to research, development and extension, Alston *et al.* (2000) also found high, but variable, returns to agricultural extension.

Nevertheless, starting in the 1990s – in the wake of structural adjustment policies and disillusionment with previous training and visit (T&V) extension – many governments gradually withdrew from funding the sector (Benson and Jafry, 2013). The T&V system was developed in the early 1970s and was promoted by the World Bank in more than 50 countries until 1998. It consisted of regular on-farm visits by field agents, who transferred technology from research institutes to contact farmers or farmers' groups acting as focal points for reaching the larger farming community. The T&V system was initially perceived as successful in a number of countries, but it did not produce results at the required scale, and had high recurrent costs (Anderson and Feder, 2007).

Recently, extension is once again the focus of attention (Anderson, 2008; Davis, 2008). After years of relative neglect,

[37] Originally, extension was largely understood as the transfer of research-based knowledge, focusing on increasing production. Today, the understanding of extension is wider and includes broader dimensions such as facilitation, learning and assistance to farmers' groups. The term "advisory services" is often used instead of extension (Davis, 2008). In line with much of the literature, this report uses the two terms interchangeably.

TABLE 7
Government and donor spending on agricultural extension and technology transfer, selected African countries

COUNTRY	NOMINAL (MILLIONS OF LCU)		REAL (MILLIONS OF CONSTANT 2006 LCU)	
	2006–07	2011–12	2006–07	2011–12
Burkina Faso	788	5 712	789	4 832
Ethiopia*	149	134	138	48
Ghana*	7.4	5.4	7.1	2.8
Kenya	3 702	7 965**	3 523	4 439**
Mali	387	461	383	390
Mozambique*	..	561	..	362
Uganda	28 023	163 572	27 159	92 512
United Republic of Tanzania	19 748	53 922	18 948	31 059

*Provisional data.
** Data refer to 2011
.. = data not available.
Notes: Numbers refer to levels of annual average spending on agricultural extension and technology transfer by donors and governments in millions of local currency units (LCU). The consumer price index (World Bank, 2013) is used to adjust nominal LCU to constant 2006 LCU.
Source: Monitoring and Analysing Food and Agricultural Policies (MAFAP) programme (FAO, 2014c).

there is now renewed recognition of the importance of disseminating and sharing agricultural knowledge among farmers. Today's agricultural extension systems have been transformed from government-driven technology transfer mechanisms to broader and more pluralistic systems of advisory services offering broader ranges of advice and involving different actors in providing it.

However, there are currently few comprehensive data on the trends and patterns of agricultural extension at the international level, regarding both expenditure and outreach to farmers. While limited data exist on public extension for some countries, achieving an overview of activities by the many non-public actors working in extension is highly problematic (Box 18).

Government spending

In many countries it is impossible to assess the scale and cost of services, even for public extension. The most recent global estimate of public expenditures on extension dates back to 1988 and put total spending at US$5 billion (Swanson, Farner and Bahal, 1988). Although estimates exist for some individual countries, the Monitoring African Food and Agricultural Policies (MAFAP) programme led by FAO in collaboration with OECD (FAO, 2014c) provides the only multi-country database that allows users to examine spending on agricultural extension. So far, MAFAP provides estimates for recent years for eight African countries: Burkina Faso, Ethiopia, Ghana, Kenya, Mali, Mozambique, Uganda and the United Republic of Tanzania. The estimates show that in most, but not all, of these countries the amount spent by governments on extension has increased in both nominal and real terms since 2006/07. The increase may partly reflect the commitment made by governments to raising spending on agriculture through the Maputo Declaration (Table 7).

Outreach

Despite their importance in providing farmers with new information on new methods and technologies, public agricultural extension and advisory services may reach fewer farmers than would be expected. The limited data available from agricultural censuses in some low- and middle-income countries suggest that only a small share of farms may interact with government extension agents.[38] In a sample of ten countries with available evidence, the share did not exceed 25 percent in any

[38] For most countries, data from agricultural censuses and household surveys relate only to interaction with public extension agents.

BOX 18
Measuring expenditure on extension and advisory services

It is increasingly difficult to measure the full extent of modern extension as it has become more decentralized, covers a broader range of areas of advice, and is often delivered by the private sector and NGOs. While compiling data on private-sector extension is next to impossible, it is more realistic to focus on government spending. Several organizations report time-series estimates of overall government expenditure on agriculture in low- and middle-income countries. These estimates include government expenditure estimates reported on the FAOSTAT database (FAO, 2013d), the International Food Policy Research Institute's (IFPRI's) Statistics of Public Expenditure for Economic Development (SPEED) database (IFPRI, 2013a) and the International Monetary Fund's (IMF's) government finance statistics (IMF, 2013). However, all of these provide estimates of spending on the agriculture sector as a whole, rather than a detailed breakdown. Providing such detail would allow users to assess spending on agricultural extension and other agricultural areas. Clearly, however, the cost and sustainability of efforts to generate such data must be considered.

As well as the MAFAP data presented in this chapter (FAO, 2014c), sources that provide disaggregated data on trends in spending on agricultural extension include the agricultural public expenditure reviews and case studies produced for individual countries by the World Bank and other development partners, including IFPRI (see for example, World Bank, 2010a; 2007a; Mogues *et al.*, 2008). Inter-country comparison using the results of such reports is prohibitively difficult because the studies do not follow a standard methodology.

Between 2009 and 2012, IFPRI, the Global Forum on Rural Advisory Services, the Inter-American Institute for Cooperation on Agriculture and FAO carried out a joint worldwide extension study. Although the study does not provide a global estimate of expenditures, it describes the financial and human resources used for agricultural extension and advisory systems at the country level, and provides information on the primary extension service providers in each country, including the primary farmers' groups they target and the degrees to which they use ICT and engage farmers in setting priorities and assessing performance.

country, and was less than 10 percent in three countries (Figure 20).

There are also indications that smaller farms are less likely to engage with agricultural extension agents than are larger ones. In a sample of household survey data from nine countries, the share of farms obtaining extension information generally increases with farm size (Figure 21), and the smallest farms are always the least likely to have access to such information. This likely reflects the poverty of many small farms and the cost of reaching them, but may be because farm income is only a small share of total household income for many small farmers (see the subsection on Multiple income sources in Chapter 2).

From India, Adhiguru, Birthal and Ganesh Kumar (2009) report that only 40 percent of farmers had obtained some kind of information on modern technology in the previous year. For large farms, the share was 54 percent, but it dropped to 38 percent for small farms. Even then, the most common sources of information were other progressive farmers and input dealers, and only 6 percent of farmers reported receiving information from government extension workers: 12 percent of large farms, and 5 percent of small ones.

While men have limited access to extension services, women farmers have even less (FAO, 2011b). There are differences between men and women farmers in the numbers of contacts with extension agents, the percentages of farmers visited by extension agents, and access to community meetings or meetings held by extension agents (Meinzen-

FIGURE 20
Shares of farms accessing information through agricultural extension, selected countries most recent year

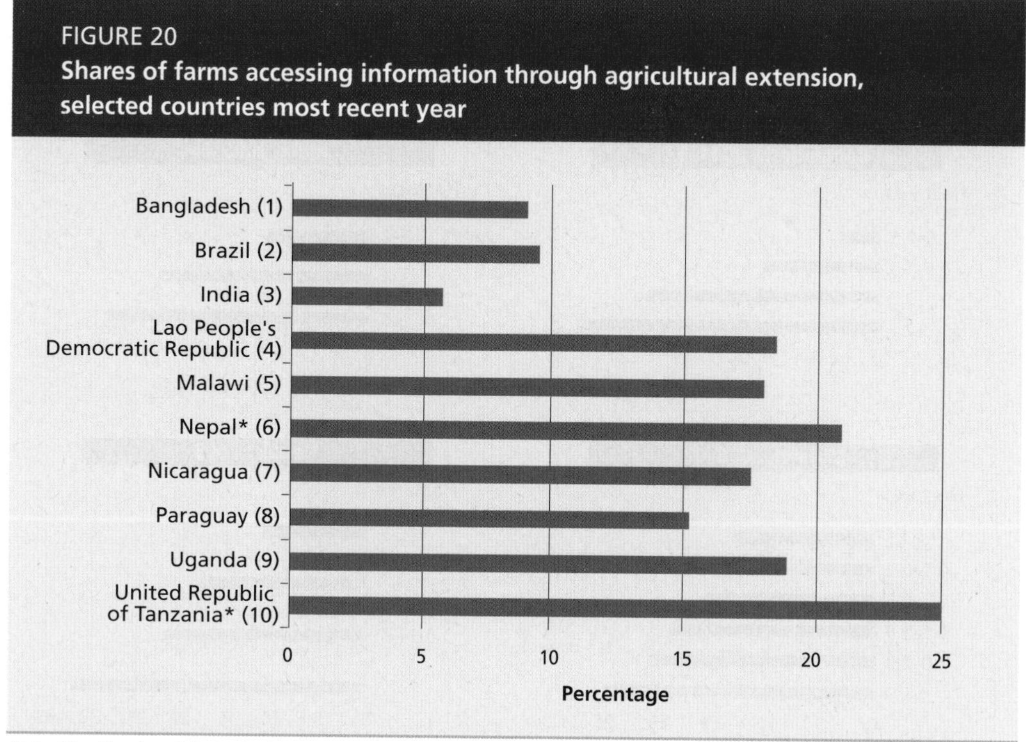

Note: *For Nepal and the United Republic of Tanzania, the shares include only farm households; non-household farm enterprises are excluded. Numbers in parentheses identify the source.
Sources: (1) IFPRI, 2013b; (2) Government of Brazil, 2009; (3) Adhiguru, Birthal and Ganesh Kumar, 2009; (4) Government of Lao People's Democratic Republic, 2012; (5) Government of Malawi, 2010; (6) FAO, 2014a; (7) Government of Nicaragua, 2012; (8) Government of Paraguay, 2009; (9) Government of Uganda, 2011; (10) FAO, 2014a.

Dick *et al.*, 2011). Extension agents often engage men farmers more than women, often partly because social norms restrict women's contacts with men extension agents. Failure to reach women at home can seriously limit their access to extension services. Time constraints and lower levels of education also prevent women from participating in certain types of extension activities unless these are specifically oriented to women. Reduced delivery of extension services to women largely reflects the lack of appropriate policies such as gender-sensitive staffing policies in extension services (Ragasa *et al.*, 2014).

Meinzen-Dick *et al.* (2011) reviewed a number of strategies that have succeeded in improving women's access to extension. These strategies include strengthening self-help groups and women's associations, affirmative action in associations and farmers' organizations, and promoting awareness of women's leadership and advocacy abilities. Other successful methods aim to recruit and train women extension agents. Intervening in public administration and the political sphere by reserving seats for women representatives in local councils or committees, creating sectoral gender focal points, and conducting gender-sensitive training for staff are other options (Meinzen-Dick *et al.*, 2011).

Extension and advisory services to meet farmers' needs

Changing paradigms for services

Increasingly, agricultural advisory bodies are called on to offer a much broader range of services than before. Globalization, economic growth and urbanization have resulted in the development of more formal market outlets, where farmers are increasingly part of value chains that extend from input suppliers to consumers. Consumers are demanding more information on the quality and safety of foods, and private-sector standards for food quality and safety are becoming more stringent. This places additional demands on producers. Environmental threats and constraints also require farmers to adapt their farming systems to sustain both productivity

FIGURE 21
Shares of farms accessing information through agricultural extension, by farm size

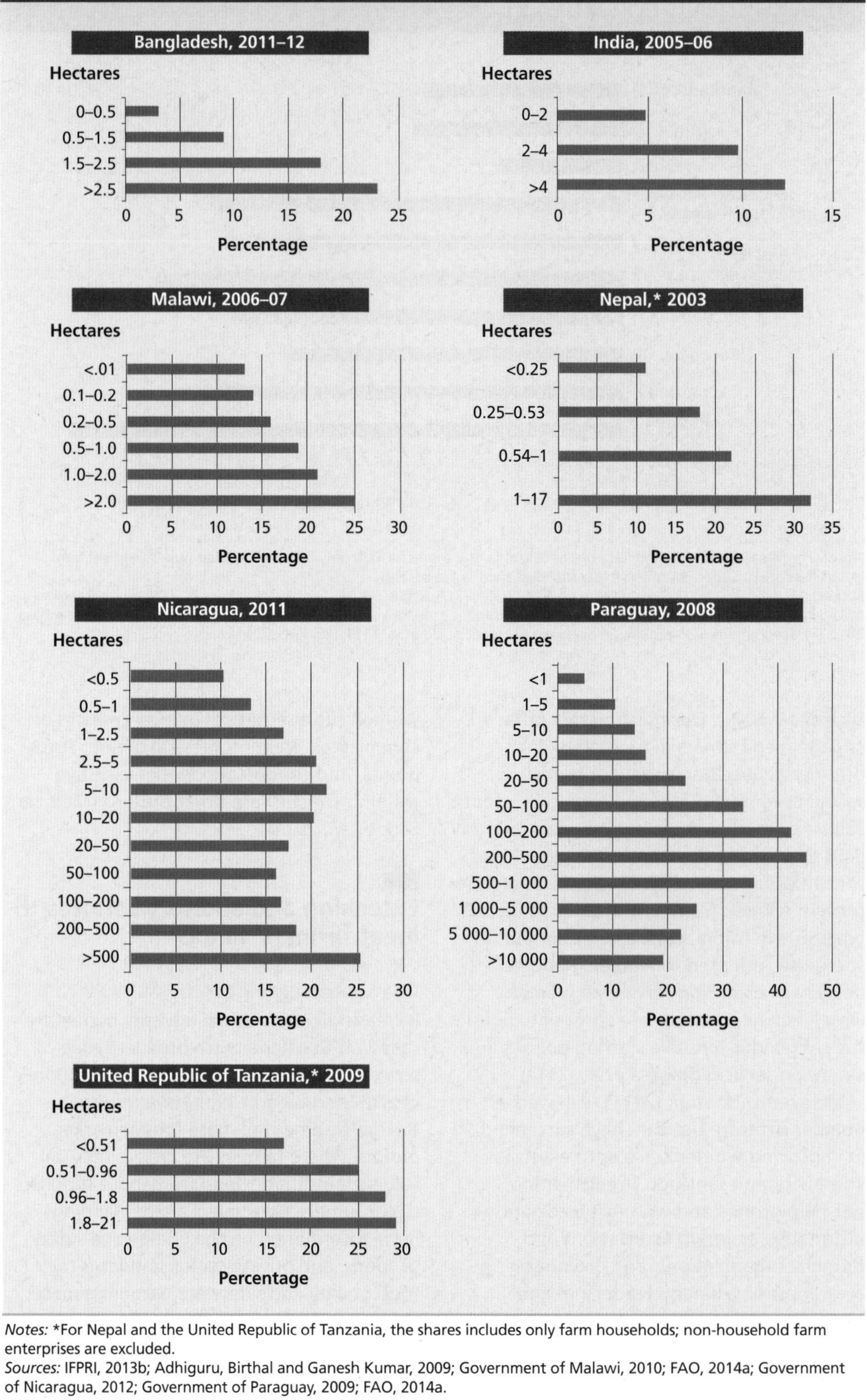

Bangladesh, 2011–12
Hectares
0–0.5
0.5–1.5
1.5–2.5
>2.5
0 5 10 15 20 25
Percentage

India, 2005–06
Hectares
0–2
2–4
>4
0 5 10 15
Percentage

Malawi, 2006–07
Hectares
<.01
0.1–0.2
0.2–0.5
0.5–1.0
1.0–2.0
>2.0
0 5 10 15 20 25 30
Percentage

Nepal,* 2003
Hectares
<0.25
0.25–0.53
0.54–1
1–17
0 5 10 15 20 25 30 35
Percentage

Nicaragua, 2011
Hectares
<0.5
0.5–1
1–2.5
2.5–5
5–10
10–20
20–50
50–100
100–200
200–500
>500
0 5 10 15 20 25 30
Percentage

Paraguay, 2008
Hectares
<1
1–5
5–10
10–20
20–50
50–100
100–200
200–500
500–1 000
1 000–5 000
5 000–10 000
>10 000
0 10 20 30 40 50
Percentage

United Republic of Tanzania,* 2009
Hectares
<0.51
0.51–0.96
0.96–1.8
1.8–21
0 5 10 15 20 25 30
Percentage

Notes: *For Nepal and the United Republic of Tanzania, the shares includes only farm households; non-household farm enterprises are excluded.
Sources: IFPRI, 2013b; Adhiguru, Birthal and Ganesh Kumar, 2009; Government of Malawi, 2010; FAO, 2014a; Government of Nicaragua, 2012; Government of Paraguay, 2009; FAO, 2014a.

and income over the long term. Diversification of sources of farm household income is another factor that broadens the demand for advisory services to cover more activities and involve different members of the farm household – men, women and youth – in different ways.

As a result, advice now covers such issues as:
- selecting the most appropriate mix of crop and livestock production;
- increasing market access;
- adding value to products and improving on-farm processing activities;
- using the most efficient production management practices;
- increasing the income and improving the welfare of farm households;
- improving management of natural resources;
- responding to climate change and other environmental threats;
- coping with risk;
- supporting producers' organizations and collaborative networks.

Advisory service must take into account the diversity of farmers' needs, which vary depending on their socio-economic conditions and the size of their household. The kind of advice that farmers require will also vary according to the quality and location of the resources under their control, their access to other physical and economic resources (e.g. credit, inputs, transportation and markets) and their technical and management skills.

Demand-responsive and participatory services

Efforts to reach small, resource-poor and marginalized farmers more effectively have included decentralization, participatory approaches and the introduction of competitive funding systems.

Decentralization can be an important way of making government-provided services more responsive to needs, but it can be expensive (Birner and Anderson, 2007). A well-documented example of decentralization of public agricultural extension is the establishment of India's Agricultural Technology Management Agency (ATMA), which is a multi-stakeholder forum that encourages collaboration among public-sector institutions, the private sector and NGOs. Features of ATMA include its use of farmers' interest groups, delivery of services by different providers, bottom-up planning, and autonomous extension system.

Participatory approaches can help make extension services more demand-driven and responsive to the needs of farmers. They can also help ensure that women's needs and specific constraints are taken into consideration, and thus contribute to removing constraints on women's productivity (FAO, 2011b). However, if participatory approaches are to succeed in this area, they must pay explicit attention to gender issues (Ragasa *et al.*, 2014). A good example of participatory approaches are Farmer Field Schools (FFS), which are community-based initiatives focusing on observation and experimentation and are now functioning in several countries around the world (Box 19).

Competitive funding systems empower farmers to experiment and discover which practices best suit them. Farmer innovation fund schemes, which may be operated by governments, NGOs or other actors, provide individual farmers, farmers' groups and other local stakeholders with small grants or loans for innovative and business initiatives selected by the recipients themselves. The schemes cover not only new technologies (on- and off-farm) and business models, but also institutional aspects such as the development of farmers' organizations (PROLINNOVA, 2012). In an extensive review of studies of innovation grants, Ton *et al.* (2013) found that the relatively few studies that assessed the impacts of innovation grants generally found positive impacts.

Delivery of advisory services by different actors

It is now widely recognized that traditional public agricultural extension cannot meet all the varying needs of different and diverse farmers and rural communities. In many countries, reforms of public-sector extension services have led to the emergence of mixed advisory systems in which services are provided by a broader range of actors, including the private sector and civil society (Sulaiman and Hall, 2002). Some governments are continuing to finance extension while contracting private firms, NGOs and farmers' organizations to provide services (Rivera and Zijp, 2002).

BOX 19
Farmer Field Schools

A Farmer Field School (FFS) is a community-based learning system in which a group of farmers studies a problem together in the field. A hands-on approach is used, with a trained facilitator – who may be an extension agent or a graduate from an FFS – leading the group through a curriculum that farmers have often chosen themselves. FFS are usually part of a government-, donor- or NGO-financed programme and sometimes work through producers' organizations. The concept was first applied to integrated pest management (IPM) in Indonesia in 1989 then spread to other Asian countries and on to many developing and transition countries. Today the focus has broadened from IPM to root crop programmes, drylands farming, livestock husbandry, market access and other activities. More than 78 countries had FFS programmes by 2005, and millions of farmers have been trained (Braun *et al.*, 2006). The FFS approach has been modified and developed to help improve farmers' access to markets through approaches such as Farm Business Schools in Asian and African countries (FAO and IFAD, 2012) and the Management Advice for Family Farms programme mainly in West Africa (Faure and Kleene, 2002). The Junior Farmer Field and Life School approach aims to empower vulnerable youth and provide them with the livelihood options and gender-sensitive skills needed for long-term food security (WFP and FAO, 2007).

While schools are widespread internationally, very little has been done to assess their performance. An impact evaluation of an FAO FFS programme in East Africa found that the income of farmers who had participated was 61 percent higher than that of non-participants, and the programme was particularly successful in improving the incomes and productivity of women, less literate and medium-scale farmers (Davis *et al.*, 2010). However, an analysis of the impact of FFS in Indonesia (Feder *et al.*, 2003) found that they did not have significant impacts on yields and pesticide use. Ricker-Gilbert *et al.* (2008) examined the cost-effectiveness of alternative methods for teaching IPM in Bangladesh, including FFS, field days and visits by extension agents. They found that FFS participants were most likely to adopt IPM but, as the schools were expensive to run, other extension methods were more cost-effective.

FFS projects and programmes have often been implemented independently of government institutions, and rely heavily on donor funding. It may be necessary to embed FFS in institutional frameworks, to expand and deepen the approach, improve quality and strengthen impact and continuity. While the FFS approach challenges the top-down extension model, its sustainability relies on the creation of an institutionally supportive environment. Key areas where such institutionalization could strengthen the FFS approach include improving the skills and quality of trainers; incorporating participatory approaches and FFS-related activities into formal education; moving from dependence on ad hoc funding from donors to more sustained financing from the public and private sectors; promoting competitive grant schemes and self-financing mechanisms; strengthening institutional support and stakeholder interactions; establishing participatory R&D methods for collaborative learning; improving the targeting of FFS participants; and standardizing procedures for monitoring and evaluation.

Joint ventures between governments and the private sector have also been created. These various formulae increase the choice of services available to farmers and are thought to strengthen incentives for improved performance (Kjær and Joughin, 2012).

The private sector

Advisory or business services may be provided by private companies or other independent service providers; many public sector-funded programmes aim to develop a cadre of such providers. In Nepal, for example, the government has set up a system of agro-vets, which are charged with supplying inputs and materials to support crop and livestock production, with the government issuing licences and providing training. Since 2003, the Swiss Agency for Development and Cooperation has operated a similar programme to promote farm enterprise development in northwest Bangladesh (Kahan, 2011).

Advisory services may also be provided by entrepreneurs selling inputs and equipment to farmers or retailers, or by the buyers of farmers' produce. In these cases, extension is often not a stand-alone activity but is provided to complement more tangible commercial services. Contract farming is often seen as a potentially effective way of delivering expertise to farmers (Box 20). Buyers generally enter into contracts with groups or individual farmers. The contracts specify the amount, quality, delivery schedule and price to be paid for produce. Farmers typically receive inputs on credit, and extension services are usually provided by the buyer to ensure that farmers meet quality standards and apply inputs appropriately (Tschirley, Minde and Boughton, 2009).

Private-sector delivery of extension services can have both advantages and disadvantages. It can facilitate the delivery of a broader array of services to different groups of farmers, but it may involve conflicts of interest, such as when private service suppliers promote specific products rather than providing more neutral information, without the possibility for farmers and their organizations to check and verify information. Private extension providers may also have no reason to be concerned about the possible negative environmental impacts of the practices they recommend, such as through excessive pesticide application or fertilizer use. The private sector has a role, but in low-income countries with generally low levels of education among farmers and without effective regulations – including environmental regulations – private-sector delivery may present pitfalls that must be recognized. A further issue may be the private sector's lack of interest in providing services to small family farms and farms in remote and marginal areas, which only public-sector engagement can serve.

Non-governmental organizations

In many parts of the world, non-profit or non-governmental organizations are active providers of advisory services, often when there is not enough commercial appeal to attract the private sector (Box 21). In rural areas that are complex or risk-prone, NGOs are frequently the main providers of

BOX 20

Contract farming and advisory service support in Sri Lanka

In 1988, Hayleys Group created Sunfrost Limited to produce semi-processed pickles and gherkins for export. Originally, the company grew the produce on a large commercial farm, but found that labour costs were prohibitive and decided to enter into contract farming arrangements with small-scale farmers. To diversify production and add value through the processing of pickles, Hayleys Group formed HJS Condiments in 1993. The company has a guaranteed buy-back system for produce in which farmers are given inputs on credit and a fixed price at which all of their produce is purchased. HJS employs a fully trained extension worker for every 100 farmers. During the farmers' first growing season, an extension agent visits them about twice a week to ensure that they are meeting quality standards; visits become less frequent in subsequent seasons. These farm visits, and training classes, are provided free to participating farmers. This arrangement has been extremely successful: by 2007, HJS Condiments was working with 8 000 small farmers and had about another 8 000 full-time employees working in producing and processing. The company accounts for 22 percent of Sri Lanka's fruit and vegetable exports.

Source: Swanson and Rajalahti, 2010.

extension services (Davis and Place, 2003; Benson and Jafry, 2013) and may provide extension advice directly or facilitate the strengthening of value chains by brokering relationships among the different actors (Kahan, 2007). NGOs have also developed methodologies for research and extension that have subsequently been adopted by the public sector (Amanor and Farrington, 1991).

NGOs have both strengths and weaknesses in providing extension services to farmers (Davis *et al.*, 2003). They tend to be participatory, demand-driven and client-centred in their approach; they have limited bureaucracy, and services are often well managed, efficient and cost-effective. On the other hand, they tend to depend on donors for funding, which can make longer-term sustainability a problem; programmes are often of short duration and geographical coverage is limited.

Farmers' groups

Farmers' organizations also play a significant role in rural advisory services. They can supply services to their members and draw on services provided from outside (Umali and Schwartz, 1994). Farmers' groups can be of various sizes and operate at different scales, and their composition may also differ. Typical groups and organizations include village-level, self-help groups; primary cooperatives; producers' associations and their federations at the regional and national levels; processing and export organizations; and national industry bodies.

Farmer-to-farmer extension relies on group-based learning, cross-visits, farmer-trainers and farmer-extension agents (World Bank, 2007a). The model originated in areas where government services were weak or non-existent. It involves self-learning and group-level cooperation, but it sometimes relies on external facilitation. Examples include the volunteer farmer-trainer approach, where farmers trained by extension staff train other farmers, host demonstration plots and share information on improved agricultural practices with their communities (Kiptot and Franzel, 2014) (Box 21).

Mixed systems

New forms of arrangement promote collaboration among the public and private sectors and civil society. Even where public financing of extension is warranted, non-State service providers are often more

BOX 21
Volunteer farmer-trainers in the East Africa Dairy Development project

The East Africa Dairy Development project is a collaborative effort among Heifer International, Technoserve, the International Livestock Research Institute, African Breeders Service Total Cattle Management and the World Agroforestry Centre. The project started in 2008 and aims to improve the incomes of 179 000 dairy farmers in Kenya, Rwanda and Uganda through improved dairy production and marketing. It uses volunteer farmers as trainers to help disseminate technologies and practices. The volunteer farmers are trained by government extension officers and host demonstration plots on which they produce seeds and train other farmers in their communities in livestock feed crops, feed conservation methods and feed formulation. The system complements, rather than substitutes for, public, NGO and private-sector extension services.

By June 2012, there were 2 676 farmer-trainers, one-third of whom were women. On average, each volunteer farmer trained 20 farmers per month and reached an average of five villages outside his/her own. They held an average of about 2.5 training sessions per month, spending about two hours per session. The most common mode of training was through farmers' groups. Women trainers were as knowledgeable as their male counterparts and reached as many farmers, even though their literacy levels were lower and they covered fewer villages.

Source: Kiptot, Franzel and Kirui, 2012.

efficient and flexible (Anderson, 2008). The public sector contracts agricultural extension in many ways, which may involve different types of public-sector agencies, local or international NGOs, universities, extension consulting firms or rural producers' organizations. These kinds of outsourcing model can be found in Mali, Mozambique, Uganda, the United Republic of Tanzania and other countries (Heemskerk, Nederlof and Wennink, 2008).

Public–private partnerships (PPPs) can support research (as seen in Chapter 4) as well as technology transfer and advisory services. Although the PPP model is considered very promising, there is still relatively little evidence of its effectiveness, partly because of its novelty. PPPs and other forms of multistakeholder collaboration also face challenges, such as in providing incentives for initiating a partnership. Cultural differences and communication difficulties among partners and stakeholders may take a long time to overcome (Spielman, Hartwich and von Grebmer, 2007). It is also important to have a strong governance framework and institutional support mechanisms to avoid restricting the range of farmers who benefit to those who can afford to pay service fees.

The National Agricultural Advisory Services (NAADS) in Uganda has generated interesting lessons regarding public–private extension services. NAADS aimed to increase agricultural production for markets by empowering farmers to demand and control agricultural advisory services. Under the programme, public extension advisers were phased out and rehired by private firms and participating NGOs, or acted as independent consultants paid by farmers. However, an analysis by IFPRI found that the evidence of "whether the NAADS program adequately induced participants to establish new enterprises or to adopt technologies and improved practices more frequently than their non-participating counterparts, seems patchy, with tenuous links ... to increased productivity and commercialization of agriculture" (Benin *et al.*, 2011). A later study attributed the limited success of NAADS partly to its over-radical approach and concluded that for complex, large-scale institutional reform programmes, gradual consensus building might work better than sweeping reforms, which risk ignoring local expertise and inviting passive resistance (Rwamigisa *et al.*, 2013).

Information and communication technology

Direct face-to-face extension services are increasingly being complemented and sometimes replaced by modern communications technology such as mobile phones, the Internet and more conventional mass media – radio, video and television (Asenso-Okyere and Mekonnen, 2012). ICT can play an important role in informing farmers and rural entrepreneurs on such issues as weather conditions (locally and in other parts of the world), input availability, dealers, financial services, market prices and buyers. Mobile phones are of particular relevance, and their use has been expanding rapidly worldwide. Cell phones have great potential for the widespread dissemination of production, marketing and management information, and for mobile banking, insurance, credit or subsidy schemes (Box 22).

In a review of studies conducted on the use of ICT for agricultural development in Africa and Asia, Asenso-Okyere and Mekonnen (2012) found that some studies showed little to no impact, while others found significant improvements in market access, on-farm income, productivity, crop diversification and environmental stewardship.

Various barriers may constrain farmers' access to ICT (Nagel, 2010; Rodrigues and Rodríguez, 2013): illiterate and older farmers are usually less likely to use computers and smartphones; the prices of broadband or mobile services are relatively high; and connectivity may not be available or its quality may be poor. Dissemination may also be limited if the content and format of the information do not match farmers' needs (Burrell and Oreglia, 2013). In a study of the benefits of providing SMS-based market and weather information to farmers in India, Fafchamps and Minten (2012) found no significant effect on prices received by farmers, crop value-added, crop losses resulting from rainstorms, or the likelihood of changing crop varieties and cultivation practices.

BOX 22
Using ICT to improve farmers' access to extension services in Uganda

In 2009, the Grameen Foundation began a partnership with Google and MTN Uganda to develop an SMS application called Farmer's Friend, which compiles agricultural information and weather forecasts into a searchable database. Farmers can text a question to the database and receive a reply via SMS (Yorke, 2009). To increase the impact of the service, the Grameen Foundation developed the Community Knowledge Worker (CKW) programme to engage local farmers in delivering information and extension services to neighbouring smallholders.

Each CKW receives a loan to obtain a "business in a box" that includes a smartphone and a solar charger. The phones are preloaded with an Android application called CKW search, which is a database that includes advice on issues such as crop pests, animal diseases, where to buy agricultural inputs, weather forecasts and marketing information (Grameen Foundation, 2013a). CKWs use the application to answer farmers' questions and encourage the use of agricultural best practices. They also conduct surveys on their phones to collect important data about smallholders and their farms. CKWs are paid to conduct the surveys, and they earn additional income from letting other people use the solar charger.

The value of CKWs is that they are respected community members who are farmers themselves and are thus are able to put the information provided through ICT services into context for other individuals. The farmers in their communities trust the CKWs, value the information they receive and are therefore more likely to apply that knowledge on their farms. The CKWs are also able to provide feedback from farmers in a two-way flow of information that helps the programme perform better.

A review conducted in 2012 showed that farmers with access to a CKW received prices that were 22 percent higher than those of farmers without access (Grameen Foundation, 2013b), and their knowledge levels rose by about 17 percent (Van Campenhout, 2012). Once human contacts were incorporated into the provision of agricultural advice through ICT services, behaviour changed and positive outcomes were achieved. The CKW programme provides a low-cost, scalable model for providing ICT-enabled extension services to poor, remote smallholders. As of 2013, the programme included more than 1 100 CKWs serving more than 176 000 farmers. It has been replicated in Colombia (Grameen Foundation, 2013a).

Developing extension and advisory services for family farmers

The role of government in mixed extension systems

In spite of the growing importance of private agricultural advisory services, for both economic and social reasons there is still a clear need for government to maintain a role in providing advice to farmers in many countries. However, it is also clear that governments can no longer be expected to act alone to meet farmers' increasingly complex needs. The challenge lies in defining the precise role of government within the framework of a mixed system of advisory services featuring many actors (Box 23).

As argued by Birner *et al.* (2009), there is no single best method for providing extension advice that responds to different needs, purposes and targets. The right approach depends on the specific policy and infrastructure environment, the capacity of potential service providers, the farming systems used, the extent of market access, and the characteristics of local communities, including their willingness and ability to cooperate. Different situations require different approaches, but to succeed, extension has to be flexible and accommodate local needs (Raabe, 2008). These include gender dimensions and the needs of women farmers (Anderson, 2008).

Governments must recognize the importance of advisory services in which different actors play different roles and provide different services to different groups of farmers. They must support and facilitate private-sector advisory services with private goods characteristics. The public sector is responsible for creating the proper conditions for private investment, such as the presence of infrastructure, education and training, as well as the right incentives and good governance.

Another important role for government is coordinating and regulating services in a pluralistic environment, including promoting coherence among services for the agriculture, pastoral, forest and fisheries sectors. Governments have a responsibility for ensuring that advisory services provided by the private sector and civil society are technically, socially and economically appropriate. Governments should provide appropriate policy formulation, analysis, quality control and regulatory functions, especially as the private sector usually has few incentives to look after the public good (Kidd *et al.*, 2000). It is particularly important to consider the possible environmental impacts of practices recommended and promoted by private extension service providers.

Governments also have a direct responsibility to provide extension and advisory services where the private sector is unlikely do so. Core areas for government involvement are sustainability and environmental concerns, the spread of crop and livestock diseases, and food safety issues (Benson and Jafry, 2013). Public concerns regarding food security and poverty eradication also call for strong public engagement in ensuring extension services.

A critical concern for governments is to ensure that services are available for small family farmers, especially in remote or marginal areas. Private extension providers are more likely to serve large commercial farms than small and sometimes remote farmers, who may be costly to reach and who may not be able to pay for services. Farmers may frequently not be aware of the benefits of extension and advice, and thus be unwilling to pay the full costs, even when able to do so.

Adequate, clearly targeted and stable public funding is necessary to ensure advisory services for small family farms and to address environmental and sustainability concerns. However, actual service delivery may be private. The best approach depends on the type of service and local circumstances. Forging effective partnerships between the public and private sectors is important, but new partnership arrangements should not be viewed as a panacea or a way for the public sector to retreat from extension. Public-sector involvement is important in ensuring that public funds are used effectively and transparently and in monitoring and supervising private-sector performance.

While recognizing the importance of public funding, governments inevitably have to take into account the trade-offs between the number and types of farmers reached and the associated costs. Providing extension services to a large number of small farmers may be very expensive without some degree of targeting of beneficiaries. When publicly funded extension services are motivated by social and equity concerns, governments must also consider whether delivering services to a large number and wide variety of farmers is more cost-effective in poverty alleviation than are possible alternatives.

However, it should not be forgotten that political economy considerations and pressure from interest groups have often tended to skew public expenditure and policies to benefit urban rather than rural dwellers and a small number of larger-scale farmers rather than a multitude of smaller farmers (see FAO, 2012b for a discussion). Governments bear a responsibility for ensuring that rural areas and smaller farms are not forgotten. Obviously, the choices made will depend on specific national and local circumstances, as well as on the government's agricultural and overall development strategies.

Gathering evidence, measuring impact and sharing experiences

There is no universally applicable type of agricultural advisory service. Birner (2009) encourages interested parties (the public, private and civil society sectors) to focus

BOX 23
Promoting innovation and competitiveness in agriculture in Peru

In the late 1990s, the Peruvian Government decided to reform its extension system and adopt an innovative approach to agricultural development. Through the Innovation and Competitiveness for Peruvian Agriculture Programme (INCAGRO), the World Bank provided a loan to establish a modern and decentralized agricultural science and technology system that is pluralistic, demand-driven and led by the private sector. Farmers played a pivotal role in managing the programme. Agriculture service providers were contracted to implement specific activities, and farmers contributed in cash and in kind to the projects. The programme generated a demand-driven market for agricultural innovation by enhancing the power of its clients – family farmers – in formulating, cofinancing, regulating, implementing, and monitoring and evaluating extension services through competitive funding mechanisms.

Over eight years of implementation, thousands of farmers demanded and received extension support. A Ministry of Agriculture study showed that 56 percent of producers had adopted new technologies, 86 percent showed productivity increases, and 77 percent were willing to pay at least part of the cost of extension services. In addition, the number of extension and research providers grew by 23 percent, and the range and quality of the services offered expanded. The same study estimated the rate of return on investments in extension at between 23 and 34 percent. The World Bank has estimated the economic rate of return at 39 percent. However, equity was a concern, as the greatest beneficiaries were male farmers and medium- to large-scale producers rather than female farmers and smaller, more disadvantaged producers.

Source: Preissing, 2012.

on creating a context-specific approach, which would include elements from existing strategies adapted to the context in which the advisory services are to be implemented.

A crucial problem facing governments and other actors in designing effective extension and advisory services is the shortage of empirical evidence to guide choices. There is little information on private- and NGO-sector investment in the provision of advisory services, or on the demand for such services from family farms. Research on the status, performance and impact of rural extension has also been limited. There has been very little comparative or *ex-post* evaluation of cases to determine whether new approaches are economically viable and whether they can be replicated and sustained in whole or in part. The often fragmented experiences of agricultural advisory services must be better understood to inform public policies.

Developing fora and mechanisms – both national and international – for the exchange of experiences and evidence on agricultural advisory services and their impact can help policy-makers and stakeholders make better decisions. At the international level, the Global Forum on Rural Advisory Services (GFRAS) represents an important effort in this direction. Its main objectives are to provide a voice for advisory services in global policy dialogues and promote improved investment in rural advisory services; to support the development and synthesis of evidence-based approaches and policies for improving the effectiveness of rural advisory services; and to strengthen actors and fora in rural advisory services through facilitating interaction and networking. Similar initiatives at the regional level include the African Forum for Agricultural Advisory Services (GFRAS, 2014), and there are also thematic networks such as the Consortium on Extension Education and Training. Further development of such efforts should be encouraged, to make advisory services more effective, inclusive and able to meet the needs of family farms.

Key messages

- Agricultural extension and advisory services are essential for closing the gap between actual and potential productivity and ensuring widespread adoption of more sustainable agricultural practices that preserve natural resources and provide crucial environmental services. Empirical evidence suggests that there are high returns to public expenditure on agricultural extension. Given the large yield gaps in many low- and middle-income countries, governments may consider increasing the priority they give to this aspect of their national innovation systems.
- Agricultural extension and advisory services can provide family farmers with information that allows them to make better and more informed choices about product mix, appropriate technologies and practices, and farm management. Too many farmers lack access to information from agricultural extension and advisory services. Smaller farmers are less likely than larger ones to have such access, and women engaged in agriculture have even less access than men.
- Different kinds of extension and advisory services, delivered by various service providers, are more likely to meet the diverse needs of different farmers: there is no standard fit. However, as in agricultural R&D, both public and private sources of extension and advisory services have important but different roles to play. Public and private roles must be clearly defined and well coordinated and regulated to foster collaboration between the public sector and different private actors. The public sector also has a responsibility to ensure that the advisory services provided by the private sector and civil society are technically sound and socially and economically appropriate.
- In spite of the growth of private advisory services, there is still a clear role for governments in the actual provision of extension services. Many types of advisory services can generate important public goods – such as lower food prices, increased sustainability and poverty reduction – that call for government intervention. Governments have a special responsibility towards small family farms, whose needs are unlikely to be met by the private sector. Governments also need to ensure provision of advisory services related to environmental sustainability and other public goods.
- Producers' organizations, cooperatives and other community-based organizations can play a central role in providing services to smallholders and helping them voice their requirements. Strengthening the capacity of family farming organizations to advocate for and provide services can help ensure more transparent and demand-driven extension and advisory services.
- There is need for more evidence on which advisory service models work best and for improved national and international information in this regard. Efforts to gather and share information about effective extension models should be promoted at both the national and international levels.

6. Promoting innovation capacity for the benefit of family farms

Previous chapters have discussed the roles of research and of extension and rural advisory services in supporting innovation on family farms. A broader challenge lies in strengthening the innovation system for the benefit of family farmers, to improve their productivity, the sustainability of their production, and their livelihoods. This chapter examines how to develop innovation capacity for family farms at different levels: individual, collective and through an enabling environment.

Developing innovation capacity

Strengthening the capacity for innovation means investing in learning and developing the skills of multiple actors in the agricultural innovation system. It also requires providing the right incentives to encourage people to put these skills into use and to develop the right attitudes and practices. The capacity to innovate can be considered as including a combination of: (i) scientific, entrepreneurial, managerial and other skills, knowledge and resources; (ii) partnerships, alliances and networks linking different sources of knowledge and different areas of social and economic activity; (iii) routines, organizational culture and traditional practices that encourage the propensity to innovate; (iv) an ability for continuously learning and using knowledge effectively; and (v) clusters of supportive policies and other incentives, governance structures and a conducive policy process (Hall and Dijkman, 2009).

The capacity for innovation can be developed in three main areas (Figure 22):

- upgrading the skills, expertise, competencies and confidence of *individuals* and organizations by building their human capital;
- improving processes within *organizations*, businesses and family farms involved in identifying and/or developing, adapting and scaling up innovations;
- creating a *policy environment* conducive to these efforts, and forging links, communication channels and networks to enable individuals and organizations to obtain and exchange new ideas and expertise for innovation.

These areas conform to the three levels of a capacity development strategy defined by the United Nations Development Programme (UNDP) and FAO (OECD, 2006; FAO, 2010b). The capacity development needs and the other interventions required will differ from country to country, depending on countries' specific circumstances. It is important that capacity development initiatives meet the needs of the recipient country (rather than of the donors) and of main actors in the national innovation system, especially family farms (Box 24).

Focusing on youth

More attention must be given to young people, who can be central to accelerating innovation in family farming. Youth may have greater awareness of new technologies, more recent education and curiosity, giving them an important role in helping their families link to broader innovation systems. Youngsters who have been employed elsewhere in the agrifood system may have experienced new ideas and technologies that they can try out with their families. Young people also often have an important role in ensuring that new information channels are used effectively. The extent to which young people perceive farming as a profession with room for innovation frequently determines whether they remain in the sector. If they see farming as dynamic and potentially profitable, they are more likely to take over their family farms.

Young people may have skills and motivation for innovating, but very often

FIGURE 22
Capacity development at different levels

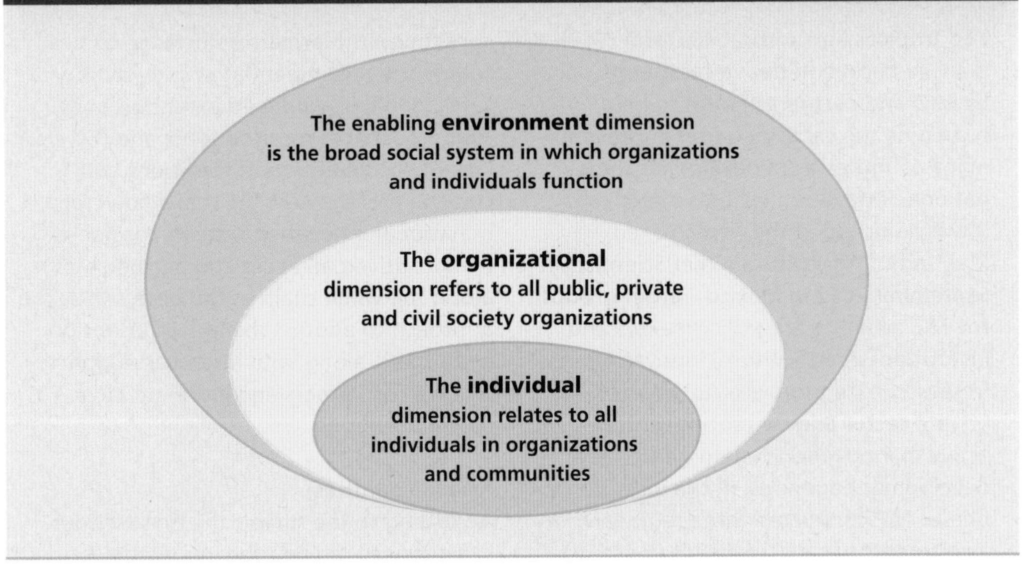

Source: FAO, 2010b.

lack access to land. Land fragmentation makes it likely that young people will inherit only small parcels of farmland, so many perceive farming as a last-resort, temporary or part-time occupation. Dysfunctional land markets reinforce existing inequalities in access to land, while well-developed rental markets can result in productivity increases of about 60 percent (Deininger, Jin and Nagarajan, 2009), thus offering the chance of an income to youngsters who would otherwise have to wait to inherit land from their relations (Proctor and Lucchesi, 2012).

Collective action through producers' and other community-based organizations presents opportunities for youngsters to earn a livelihood from agriculture even if they have not yet inherited land. Some young people use producers' organizations as a base for offering services such as processing, collection or transport. Others, with higher education levels, are able to find employment in the middle management of NGOs. It has been noted that the social networking associated with collective action can generally help to make smallholder farming more attractive to rural youth (Proctor and Lucchesi, 2012). It is also recognized that ICT is changing the role of young people in societal development (Shah and Jansen, 2011).

Developing individual capacities

Education and training represent an investment in people and are probably the most important way to develop people's skills and competencies for innovation, whether they are farmers, service providers, researchers or policy-makers. Farmers need to attain more advanced levels of education to make use of new ICT-based information sources and technical advice and to respond to new market opportunities and environmental change. Extension staff need both an up-to-date understanding of the topics on which they provide advice and the ability to communicate and interact with other actors. Academics need to be up to date with cutting-edge science and able to address the challenges faced by family farmers when these are relevant to their research agendas.

Basic education is the most fundamental part of human resources development, not only as a universal human right, but also as the foundation for improving agricultural productivity and farm incomes. Basic education in rural areas has a significant positive impact on agricultural productivity (Reimers *et al.*, 2013). Basic education can significantly improve the efficacy of training and extension by

BOX 24
Assessing capacity development needs: the Tropical Agricultural Platform

The Tropical Agricultural Platform (TAP) is a G20-backed initiative facilitated by FAO and partners. It aims to help overcome the capacity gap that prevents many countries from developing their national innovation systems effectively. It was launched at the first G20-led Meeting of Agricultural Chief Scientists in September 2012 in Mexico. Target groups for TAP activities are policy-makers and institutions in agricultural innovation (research, extension, education, etc.), private-sector and civil society entities active in innovation systems, and relevant development agencies. In the inception phase, TAP conducted three regional assessments of capacity needs in groups of countries in Africa, Central America and Asia, based on surveys of actors in agricultural innovation systems (see FAO, 2013f for a summary of results). The surveys identified major challenges, issues and gaps – capacity development needs – in each region.

Africa (15 countries)
In Africa, the survey pointed to the need to "repackage smallholder agriculture as a business instead of sticking to the current peasant nature of agricultural systems". Key challenges for innovation include: (i) resource endowment – limited access to innovation finance, high costs of new technology and equipment, lack of farmer training centres and lack of communications infrastructure; (ii) attitudes and mindsets – inadequate participation in innovation meetings, and negative cultural values regarding new varieties and technologies; (iii) environmental challenges – desertification and climate change; and (iv) access to markets for value-added products.

Central America (7 countries)
Major concerns revealed by the survey are: (i) the limited adoption of innovations, partly because proposed innovations may be unsuitable to agro-ecological, climate and weather conditions; (ii) farmers' reluctance to follow the recommendations of advisory services; (iii) ill-equipped extension and support services for producers; and (iv) lack of consideration of traditions and cultural preferences. The surveyed actors in national innovation systems considered market-driven alliances and partnerships along the value chain as the best approach to address the lack of adoption by farmers, along with improved support services for farmers and more effective communications.

Asia (5 countries)
According to the survey, the most serious constraint to making the innovation system more effective and farmer-oriented is the lack of facilitating policies to promote capacity development. There is also a perceived lack of private-sector involvement in the agricultural economy, with a possible crowding-out effect from donor and public-sector activities. Key players in innovation enhancement include public advisory and extension services, national research institutions and the domestic private sector. Technologies such as biotechnology and information technologies are perceived as having positive environmental, economic and social impacts. Institutional and management innovations – such as enabling policies for extension, technology, microfinance and business – could help address the challenges facing national innovation systems. Private-public partnerships could be encouraged by government incentives (matching grants, tax credits, etc.), cooperation platforms and national marketing boards.

Source: FAO, 2013f.

facilitating: (i) enhanced productivity of inputs, including labour; (ii) reduced costs of acquiring and using information about technology that can increase productivity; and (iii) entrepreneurship and responses to changing market conditions and technological developments (Schultz, 1964). Special attention must be given to women, as gender differences in education at all levels are pervasive and well documented. Although educational gender gaps have tended to narrow, most significantly in Latin America, large gaps remain in South Asia and sub-Saharan Africa. Affirmative action to increase school attendance by girls can play a role in empowering the next generation of women while creating a critical mass of educated farmers and a pool of potential women actors in the innovation system (Ragasa *et al.*, 2014). The incidence of child labour in agriculture can limit children's access to basic education, and thus their ability to build the human capital needed to act as future innovators.

In addition to basic education, agricultural universities, vocational and technical colleges and farmer training centres also play a role in creating the human capital needed to modernize the sector. Agricultural education and training raises agricultural productivity by developing producers' capacities and generating human capital for research and advisory services. The development of agricultural education and training has been an integral part of the strategies of countries that have prioritized agricultural growth, such as Brazil, India and Malaysia (World Bank, 2007a).

Despite the unquestioned importance of developing human resources, the agricultural education sector has not generally benefited from adequate investment. In many developing countries, agricultural training at high schools and universities has been caught in a vicious circle of low investment leading to declines in the quality of education, which in turn send enrolment rates down (Beintema *et al.*, 2012). According to a FAO report, "training programmes are not often appreciated by public-sector agencies and donors and, although a demand may exist, clients are reluctant to pay for such training. Training tends to be perceived as a 'black hole' consuming resources and infrequently offering evidence of impact. Some of the criticism is due to the not so apparent connection between training, skills development and impact" (FAO, 2008b).

Studies on the content of training suggest that failure often results from weak design and organization of curricula (Kahan, 2007). Many training courses for advisory service practitioners are too general (relying on standardized material), theoretical and supply-driven, and the quality of trainers and training delivery is often poor. Training courses also tend to be treated as single events, with inadequate follow-up. If agricultural production on small family farms is to become more market-oriented, much of the content of education and extension should be refocused to cover new technical areas such as farm management, agribusiness development, value addition and marketing (Kahan, 2007; Rivera, 2011). Evidence also suggests that training should be largely experience-based, practical and problem-oriented (Kilpatrick, 2005; Kahan, 2007), and should simulate the challenges that farmers face in the more competitive agricultural environment.

Training in innovation brokerage is an important part of skills development for advisory service practitioners, enabling them to facilitate and promote innovation that benefits family farms. Extension agents have often been trained to consider themselves as "experts" and are unaccustomed to facilitating the learning and innovation processes of others. New skills in communication, dialogue and conflict management need to be developed within public extension organizations and among private, NGO and farmer-led advisory service providers (Leeuwis and Van den Ban, 2004).

Investment is needed in developing new tertiary-level curricula that foster capacity to deal with new problems and challenges while ensuring that students gain specialized skills to address the productivity constraints of family farmers. In addition to "hard" skills in cutting-edge sectors such as biotechnology, food safety, agro-biodiversity, agribusiness and information systems, there is also need for "soft" skills such as communication and facilitation, which are essential in multidisciplinary and multistakeholder work settings (FARA, 2005; Posthumus, Martin and Chancellor, 2012).

To improve the relevance and effectiveness of education, it is also important to bring education institutes into closer, more productive relationships with other actors in the agriculture sector and the wider economy (World Bank, 2007b). With stronger linkages among education institutes, national extension systems and other stakeholders, the education and research agendas can be tailored to the needs of different user communities (Davis, Ekboir and Spielman, 2008). Focusing on Africa, Spielman and Birner (2008) call for reforms in agricultural education and training to strengthen the innovative capabilities of agricultural organizations and professionals. According to the authors, it is particularly important to align the mandates of agricultural education and training organizations with national development goals by designing education programmes that are strategically matched with the different needs of society and linked to institutions and individuals beyond the formal agricultural education system. These reforms should also include development of incentives for forging stronger links among the agricultural education and training system, other knowledge sources, the private sector and farmers (Spielman and Birner, 2008; Davis, Ekboir and Spielman, 2008).

The capacities of people at lower academic levels, such as graduates from technical colleges and agricultural schools, are also valuable in making technical skills available to the farming community. The importance of agricultural education at these different levels has often been underestimated, and there is a persistent shortage of skilled technicians in knowledge-based commercial agriculture, with its emphasis on value addition and marketing (World Bank, 2010b).

The low level of training of a large proportion of extension workers is a particular issue for many developing countries. However, as the number of middle- and college-level agriculture graduates increases, the older high school-trained extension workers can be gradually replaced. This is already happening in many countries of Asia, Latin America and the Near East (FAO, 1995).

The major challenges in agricultural education and training that face developing countries can be summarized as inadequate institutional capacity; relatively low levels of public and private support to agricultural education; and limited resources and experience to cope with new areas of training in agriculture: environment and natural resources management, biotechnology, farming systems management and agribusiness. Building a productive and financially sustainable education system requires sustained political support for investments in agricultural education and training to develop a network of core institutions (Eicher, 2006). Long-term commitment is necessary to build up the required human capital within the innovation system, while recognizing that the system needs to be dynamic to match the supply of education and training with demand (World Bank, 2007b).

Developing organizational capacity

The ability of small family farmers to arrange collective action through producers' and other community-based organizations is crucial to their capacity to innovate. It allows them to obtain access to input and output markets, to participate in value chains, and to engage effectively with other actors in the innovation system, such as research institutes and private and public advisory services. Without the capacity to organize themselves, family farmers have little influence over the social, economic and political processes affecting them.

Farmers' organizations can facilitate access to knowledge sources, inputs and markets. However, their contribution to agricultural innovation varies, depending on their mission, background, assets and networks. Farmers' organizations typically contribute to so-called support functions within the agricultural innovation system, such as input supply, credit and savings schemes, and marketing of produce. Contributions to research and extension are less common, but farmers' organizations can develop the capacity to demand services from other actors within the agricultural innovation systems (Heemskerk, Nederlof and Wennink, 2008; Wennink and Heemskerk, 2006).

In a review of good practices for building innovative rural institutions, FAO and IFAD (2012) discuss four different domains in which rural organizations can support small farmers: enhancing access to and management of natural resources; facilitating access to input and output markets; improving access to information and knowledge; and enabling small producers to engage in policy-making. All areas are important in allowing small family farms to innovate successfully. Collective action for access to knowledge and information can help small farmers create linkages to service providers, share experiences and receive training to develop both their technical and managerial capacities. FAO and IFAD (2012) provide case studies of successful arrangements in various domains of information sharing involving producers' organizations. These arrangements include strengthening the linkages between research and the needs of small producers, improving technical and managerial competencies, and promoting the use of new communication technologies.

Studies have shown severe gender biases in most farmers' organizations, natural resource management groups and other community-based organizations; these biases not only disempower women, but also reduce the effectiveness of the institutions (Pandolfelli, Meinzen-Dick and Dohrn, 2008). Overcoming gender bias and the exclusion of women from positions of responsibility requires an understanding of the different motivations and incentives facing men and women when engaging in collective action. Proactive measures are needed to promote the effective participation of women in mixed producers' organizations and cooperatives by encouraging women's leadership. For example, cooperatives have transformed the Indian dairy industry by aggregating the production of millions of men and women through a three-tiered collection system to which even the smallest producers can contribute (Narayan and Kapoor, 2008). Measures to support existing "women-only" producers' organizations have also proved valuable (FAO/IFAD, 2012).

Producers' organizations can have a significant impact through the diffusion of ideas and the development of capacities, but effective organizations can generally not be created through action from outside. Collective action is best generated from within. Producers' organizations created under pressure from projects or decentralization have rarely been sustainable. Externally induced collective action using blueprints for establishing new types of committees and platforms can ultimately even damage a community's pre-existing social capital (Vollan, 2012).

Greater understanding is needed of how to foster a culture of collective action and facilitate the creation of innovation-oriented producers' organizations. In addition, organizational capacity should be strengthened across the innovation system, not only at the level of farmers. Developing innovation capacity requires that all actors and organizations within the public (e.g. research, extension, education) and private sectors invest in becoming "learning organizations". Research and development organizations and educational and training institutes – as parts of the innovation system – may also need to introduce and develop new processes to promote knowledge management and sharing.

Building an enabling environment

While developing human and organizational capacities is important, alone it is not enough to foster innovation. A well-functioning enabling environment – including policies and rules that govern the mandates and operations of research and extension organizations and their engagement with other actors in the system – is vital for individuals and organizations to perform more effectively. Infrastructure is another core component of the enabling environment for innovation, including infrastructure to facilitate market access (e.g. roads and storage facilities), infrastructure for energy and water, and financial infrastructure. The enabling environment creates the conditions necessary for innovation to occur within society and is essential to effective innovation at the international, national and local levels (Rajalahti, Janssen and Pehu, 2008).

The State of Food and Agriculture 2012: Investing in agriculture for a better future (FAO, 2012b), discussed the enabling environment required to foster private investment in agriculture, including by smallholders (Box 25). Most of this discussion is equally relevant to innovation by farmers

BOX 25
Promoting investments in agriculture

The State of Food and Agriculture 2012: Investing in agriculture for a better future argued that more and better investments are needed in agriculture. It showed that farmers are the largest investors in developing-country agriculture and emphasized that they must therefore be central to any strategy aimed at promoting agricultural investments. The report also presented evidence showing how public resources can be used more effectively to catalyse private investment and how to channel public and private resources towards more socially beneficial outcomes. Two key issues discussed in the report were how to create a general investment climate conducive to private investments in agriculture, and how to help smallholders overcome the specific constraints they face to investing.

Creating a conducive investment climate

Farmers' investment decisions are directly influenced by the investment climate within which they operate. Farmers in many low- and middle-income countries often face an unfavourable environment and weak incentives to invest in agriculture. While many farmers invest even in unsupportive investment climates (because they may have few alternatives), evidence shows that they invest more in the presence of a conducive investment climate.

A conducive investment climate depends on markets and governments. Markets generate price incentives that signal to farmers and other private entrepreneurs when and where opportunities exist for making profitable investments. Governments can influence the market incentives for investment in agriculture relative to other sectors through support or taxation of the agriculture sector, exchange rates and trade policies. Governments are also responsible for creating the legal, policy and institutional environment that enables private investors to respond to market opportunities in socially responsible ways. Many elements of a good general investment climate are equally or more important for agriculture: good governance, macroeconomic stability, transparent and stable trade policies, effective market institutions, and respect for property rights. Ensuring an appropriate framework for investment in agriculture also requires that environmental costs and benefits are incorporated into the economic incentives for investors in agriculture and that mechanisms facilitating the transition to sustainable production systems are established.

Helping smallholders overcome challenges to investment

Smallholders often face specific constraints to investment, including extreme poverty, weak property rights, poor access to markets and financial services, vulnerability to shocks, and limited ability to tolerate risk. Ensuring a level playing-field between smallholders and larger investors is important for reasons of both equity and economic efficiency, particularly for women engaged in agriculture, who often encounter even more severe constraints.

Effective and inclusive producers' organizations can enable smallholders to overcome some of the constraints relating to access to markets, natural resources and financial services. Social transfers and safety net schemes can also be instrumental in overcoming two of the most severe constraints faced by poor smallholders: lack of savings or access to credit, and lack of insurance against risk. Such mechanisms can allow poor smallholders and rural households to build assets and overcome poverty traps, but households' choice of assets (human, physical, natural or financial capital) and activities (farming or non-farm activities) will depend on the overall incentive structure as well as the households' individual circumstances.

Source: FAO, 2012b.

and will not be repeated here. The following subsections discuss two broad issues of particular significance to the development of innovation capacity: the forging of networks and partnerships, and the need for a policy framework supporting agricultural innovation.

Networks and partnerships for innovation

Innovation at the farm level is occurring increasingly within network-like settings where farmers interact and learn from other farmers, input suppliers, traders, advisory service providers, etc. Innovation does not take place in isolation. One challenge is therefore to identify effective coordination mechanisms and systems that can facilitate interaction and coherence among actors in value chains and innovation systems. Two mechanisms being discussed and promoted are innovation brokers and innovation platforms.

A decisive factor for successful innovation is facilitation of knowledge sharing, which is the role played by *innovation brokers*. An innovation broker is a person or organization that can help overcome shortages of information about what potential partners can offer, and thus bring stakeholders together and create networks and linkages among them (Klerkx and Gildemacher, 2012). Key functions of innovation brokers typically include analysing and articulating demand, organizing networks, and facilitating interaction. Innovation brokers can come from the public, private or third sectors: national or international NGOs, international donor agencies, farmers' and industry organizations, research and extension organizations, specialist third-party organizations, government organizations, ICT-based brokers, etc. (Klerkx, Hall and Leeuwis, 2009).

Innovation platforms have been promoted as a practical approach for putting the agricultural innovation system into action (Klerkx, Aarts and Leeuwis, 2010; Nederlof, Wongtschowski and van der Lee, 2011). The platforms are mechanisms that help stakeholders interact in a concerted manner. They can provide a space for information exchange, negotiation, planning and action, and can bring together stakeholders at different levels in the innovation system to work towards a common goal. Applied in natural resource management as a way of solving problems that require collective action (Adekunle and Fatunabi, 2012), innovation platforms have also been successfully used for this purpose in agriculture.

Diverse membership is a key component of a successful platform. As Thiele *et al.* (2009) point out, a producers' organization is not a platform, because it represents and works for the interests of only producers. Similarly, Farmer Field Schools are not necessarily platforms. While they may have linkages to other stakeholders, FFS do not typically include other types of actors, such as researchers or traders; instead, they focus on developing farmers' individual and organizational capacities. However, an FFS can lead into a platform if the farm group involved connects with other stakeholders to address systemic issues.

Innovation platforms can encourage face-to-face dialogue, build trust and provide space for stakeholders to collaborate and innovate. Platforms are often set up at the local level to improve the efficiency of a specific value chain. They can be particularly useful in engaging the private sector in targeted innovation processes. Platforms at the national or regional levels often set the agenda for agricultural development and enable farmers, through their representatives, to be involved in policy-making (Box 26).

Governments can support the establishment and functioning of these networks and platforms, for instance by convening meetings with key actors at the country level to influence regional political, policy and economic bodies. Networks should be designed not only to provide technical information but also to facilitate the flow of other types of information (e.g. commercial or managerial) among a wide range of actors. It is important that platforms also involve the private sector to integrate it into the innovation system (OECD, 2013).

At the global and regional levels, there is a similar need to strengthen existing networks and establish new ones to foster collaboration and coordination in designing and sharing innovations. The Global Forum on Agricultural Research (GFAR), the Global Conference on Agricultural Research for Development (GCARD), the Global Forum for Rural Advisory Services (GFRAS) and the Tropical Agricultural Platform (TAP)

BOX 26
Innovation platforms from Africa

Maize and legumes, Nigeria
This innovation platform brought together farmers, researchers, capacity-building organizations, national extension services, the private sector and local government. Together, participants set up training programmes and joint experiments and supported farmers' organizations. The platform triggered the development of an apex farmers' organization to enable direct negotiations between farmers and private companies. Achievements of the platform included improved maize-legume production systems; facilitation of mutual learning processes among platform members; participatory experimentation with farmers; interorganizational coordination to support change processes; development of an apex farmers' organization and new networks; and training of lead farmers in pilot villages to disseminate new practices.

Oil-palm, Ghana
This innovation platform was organized on two levels. At the local level, experiments took place with small-scale processors to improve their practices. Findings fed into the higher-level platform, which lobbied for policy changes at the national level and for change in the practices of oil-palm producers and processors. Achievements of the platform included increased eagerness of stakeholders to experiment and improve their knowledge; inclusion of women members in district assemblies; engagement of district assemblies in discussions on small-scale processing; and more attention to small-scale processing activities from large organizations (e.g. the Oil Palm Research Institute, the Ministry of Agriculture).

Cowpeas and soybeans, Nigeria
The objective of this innovation platform was to address selected practical problems in the soybean and cowpea value chains. As a group, platform members (mainly women farmers) were able to meet banks, policy-makers and other stakeholders that were previously not accessible to them. Achievements of the platform included improvements in seed distribution; training of farmers in cowpea storage and management of fodder stockpiles; and presentation of a study on national policies to policy-makers.

Soybean, Ghana
A soybean cluster, consisting of stakeholders active in a local soybean value chain supported the formation of farmers' groups and their involvement in the development of soybean varieties and technologies. It also established an important forum for stakeholders in the soybean sector to meet and negotiate trade and marketing opportunities. Achievements included all members' acknowledgement of learning from working together (about technologies, operating as a value chain); access to credit; intensification of soybean production; increased membership in the platform because of the crop's popularity and the association-building skills of platform members; and increased interest in commercial production from subsistence-oriented farmers.

Source: Nederlof, Wongtschowski and van der Lee, 2011.

are examples of initiatives that involve broad groups of stakeholders. It is also important to build a publicly led system for technology sharing at the global level and networks of international research and application centres to improve the diffusion of appropriate technologies for sustainable productivity (United Nations, 2011).

Policies for fostering innovation

Governments have a lead role in setting clear objectives for the agriculture sector and formulating policies that promote agricultural innovation. Policies promoting agricultural innovation can either be developed separately for the agriculture sector or be embedded in an umbrella national innovation strategy (Anandajayasekeram, 2011). Governments in emerging economies increasingly recognize that a purely sectoral approach is not sufficient, and tend to see their agricultural innovation systems and associated policies as part of a larger national strategy for innovation affecting all sectors (Tropical Agriculture Platform, 2013). In addition, more than in other sectors, political interests in agriculture tend to benefit from maintaining the existing situation; embedding policies to promote innovation in agriculture within overall strategies can help overcome this strong resistance to change (FAO, 2013f).

A national innovation policy provides guidance on how to coordinate a wide spectrum of policy domains – science and technology, education, and economic, industrial, infrastructure and taxation, among others – to create an environment that stimulates innovation (Roseboom, 2012). Strategies need to take into account the range of policies and regulations that affect the capacity of all sectors to create and adopt innovation, and systems of incentives or disincentives to foster innovation are needed. Eliminating the main impediments to innovation involves ensuring a stable macroeconomic environment and open and well-functioning markets. It also requires setting appropriate regulations transparently and fostering human capital. Other measures include policies for health, education and infrastructure.

Policy coherence is essential in improving the performance of an innovation system that supports family farming. A national innovation policy needs to define the roles of the different contributing ministries and other stakeholders in the system and to set priorities for public investment across sectors. Coordination at the local, national, regional and international levels is crucial, given the growing number of actors in the innovation system and the increasing complexity of international challenges.

The high-level innovation councils found in some OECD countries can play important roles in setting priorities and agendas and acting as an overall policy coordination platform (Finland and the Republic of Korea are examples of countries with such bodies). However, their tasks must be well defined (Hazell and Hess, 2010). The composition of an innovation council needs to be considered in the light of the strategic tasks to be implemented, and must include representatives of the private sector, NGOs and smallholders. Where innovation strategies are incorporated in agriculture-related ministries, a higher-level entity is sometimes set up to coordinate relevant policies among the appropriate ministries (Roseboom, 2012).

The regulatory environment can strongly affect innovation among family farmers by setting standards, reducing risks, decreasing administrative burdens and responding to market failures. Inappropriate regulations can delay technological progress and transfer, and impose excessive transaction costs on farmers' and other organizations. The regulatory environment for fostering innovation in family farming encompasses such issues as access to markets, particularly where markets are weak; access to land where land markets and security of tenure are absent; laws pertaining to contracts, to promote contract farming; intellectual property rights; health and food safety; biosafety and environmental regulations; and the legal arrangements for farmers' organizations (OECD, 2013).

In a survey of peer-reviewed research on the adoption and impact of transgenic crops in developing countries, Raney (2006) concluded that institutional factors – such as national agricultural research capacity, environmental and food safety regulations, intellectual property rights and agricultural input markets – are at least as important as the technology itself in determining the level and distribution of economic benefits to farmers and other actors. In China, for example, successful adoption of insect-resistant cotton depended on the strength of the highly developed public agricultural research system, and was found to be decidedly pro-poor, as proportional income gains on small and medium-sized farms were more than twice those on the largest farms. In contrast, in Argentina, strict enforcement of intellectual property rights for insect-resistant cotton, and the high costs of seeds, limited economic benefits and thus adoption. However, unpatented, transgenic, herbicide-tolerant soybeans were widely adopted, leading to an estimated increase of 10 percent in total factor productivity. Evidence from South

Africa underlines the role of local institutions in the adoption of new crop varieties; several studies found positive and pro-poor impacts for smallholder farmers in areas where a local cooperative provided insect-resistant cottonseed on credit, along with technical advice. However, this initiative was successful only because the cooperative ran the only cotton gin in the area so could ensure that loans to farmers were recovered; when another cotton gin opened in the region, the cooperative was no longer assured of debt recovery and ceased providing insect-resistant cottonseed on credit.

Policies, public investments and the regulatory environment have significant implications for the ways in which agricultural products are produced and reach domestic and foreign markets, for promoting private investment in agricultural R&D, and for fostering innovation and the use of more sustainable agricultural practices by family farmers (Roseboom, 2012). Policies may also determine which stakeholders benefit most from innovation, by emphasizing large or small farms, commercialization rather than food security, or enterprises dominated by men rather than women. For example, if policies fail to address the challenges that women face in securing land tenure, women may be less interested in investing in more intensified production. It is up to governments to make the right choices based on their development objectives and policy priorities (Box 27).

A major issue is ensuring that policies to support innovation take into consideration and address the concerns of small family farms. Policy-makers are often not fully aware of the challenges faced by family farmers, or of family farmers' role in agricultural growth and sustainable development. The pervasive and persistent influence of elite groups has been identified as the primary obstacle to reforms in research and extension systems (see, for example, Poulton and Kanyinga, 2013). This undue influence can partially be attributed to small farmers' limited ability to make their voices heard, and/or a failure to ensure broad consultative structures that include family farmers. As a result, public policies often favour larger, commercial farmers over smaller family farms.

Rural institutions, particularly powerful producers' organizations, can defend the interests of family farmers by enhancing their participation in formulating and implementing the policies, programmes and projects that concern them (Bienabe and Le Coq, 2004). The challenge for family farmers is to build a collective voice to ensure that their concerns are taken into consideration in policy formulation and national development planning.

The participation of small-scale producers' organizations in the design of public policies and in public–private sector dialogue helps to guarantee that public policy-makers listen to the voice of rural people. Participatory mechanisms reveal people's needs and provide quality information to governments and public institutions, helping them to design appropriate and effective agricultural and rural development policies. To ensure that the voices of all farmers are heard, it is indispensable that women are actively engaged in these processes.

In recent years, organizations of farmers and other producers in Latin America, Asia and Africa have established regional networks to strengthen their capacities and influence national and regional policies. These fora include the Confederation of Family Farmer Producer Organisations (COPROFAM), the Asian Farmers' Association for Sustainable Rural Development (AFA), the Network of Peasant Farmers' and Agricultural Producers' Organizations of West Africa (ROPPA) and the East African Farmers' Federation (EAFF). They allow family farmers to participate in decision-making through deliberative processes with governments and other actors. However, family farmers still need to strengthen their capacities to participate in and influence policy dialogue and decisions, to create an enabling environment that is more favourable to them and their needs.

Measuring, learning and scaling up

Many examples of good practices in innovation among family farmers are from pilot projects (Box 28). There is not yet enough empirical evidence on how these practices affect smallholder productivity and income and on the potential for replicating and adapting them. One reason for this

BOX 27
Agricultural innovation in sub-Saharan Africa

The Forum for Agricultural Research in Africa (FARA) reviewed 21 case studies of innovation approaches across sub-Saharan Africa. The objective was to draw lessons on the usefulness of these approaches in guiding research agendas to improve food security and nutrition, reduce poverty and generate cash incomes for resource-poor farmers. FARA concluded that:

> *The case studies demonstrated that successful multiple stakeholder approaches are dependent on a wide range of facilitating and inhibiting factors. Enabling public policies and regulations, including deregulation of markets, whilst ensuring competition and compliance with minimum standards often provide a solid foundation. The creation of a network of stakeholder groups drawn from both public and private sectors is a prerequisite. Such groups need to have the capacity, capability and willingness to interact and work together in an environment that encourages cooperation, builds trust and establishes a common vision for the future. The establishment and participation of effective and representative farmer organisations able and willing to communicate with members is vital. In most cases this required support and capacity development.*
>
> *Clearly, improved infrastructure, particularly roads, communication and power provide the basis for ensuring inputs can be made available at affordable prices and outputs delivered to market. This was often a precursor in seeking opportunity to add value along market chains. Easy and timely access to inputs, including finance, is crucial and needs to be based on effective and competitive marketing, whether domestic or export, and to address social and environmental concerns.*
>
> *Although research can be an important component, it is often not the central one, and in the early stages, interventions to develop capacity, access and use existing knowledge, and foster learning are required.*

Source: Adekunle *et al.*, 2012.

shortage is that innovation processes are slow, so their impact may only be apparent after a decade or more (Triomphe *et al.*, 2013). In addition, the diversity of agriculture, combined with the complexity of development, has significant implications for scaling up. What works in one setting cannot necessarily be replicated elsewhere with the same results. Innovation is a dynamic and uncertain process that cannot be predicted (Klerkx and Gildemacher, 2012) or easily attributed to individual actors or actions (Ekboir, 2003).

A defining feature of agriculture is the enormous differences among different locations in terms of agro-ecological conditions, production and market opportunities, services, infrastructure, human capacities, culture, etc. The constellations of local stakeholders involved in innovation processes also vary, as do the types and levels of access to knowledge from outside the location. A technology or institutional change process that worked well in one place will not necessarily work well in another, and a multistakeholder effort along a value chain that functions today may need to change tomorrow, depending on the market.

For scaling up, the competencies of researchers, farmers, extension staff, development planners and policy-makers need to be developed, and systems of learning and knowledge sharing designed. Indicators for measuring the outcomes of capacity development are also needed. Scale-up requires monitoring and evaluation (M&E) systems to process the flow of information from new and often very local experiences. M&E may focus on monitoring quantitative aspects, such as farmers' adoption rates or the extent to which farmers adapt technologies to their own situations, but it also involves assessing qualitative institutional changes, including policies,

BOX 28
Experiences of agricultural innovation in Africa

As part of the European-funded Joint Learning in and about Innovation Systems in African Agriculture (JOLISAA) project, an inventory was made of agricultural innovation experiences in Benin, Kenya and South Africa. The objective was to assess multistakeholder agricultural innovation processes involving smallholders. The complete inventory includes 57 documented cases covering a wide variety of experiences.

A number of trends were detected and can be summarized as follows:

- *Market-driven innovation* may take place through the emergence of new value-chain arrangements, or when producers take into account the demands or standards of consumers or industry. The emergence of market-driven innovation was identified in many cases across the three countries, usually combining elements of technical innovation with organizational or institutional ones.
- *Lead and active stakeholders* varied, depending on the specific case and the phase of the innovation process. For instance, researchers, an NGO or an R&D project might have been very active in the early stages (conducting diagnoses and on-farm experimentation, providing capacity development, etc.), while farmers and their organizations or business stakeholders became more active later. Researchers did not necessarily play a leading role or initiate innovation in many of the inventory cases, as ideas and initiatives came from different sources, including farmers themselves.
- *Interactions among stakeholders* were rather informal in some cases; in others they took place under the umbrella of an R&D project and/or multistakeholder platforms, especially when a common resource (e.g. a mangrove, irrigation scheme or forest) needed to be managed (Hounkonnou *et al.*, 2012). In many cases, one of the actors (typically a research institute or an NGO) acted as the intermediary or innovation broker to facilitate interactions among stakeholders.
- Most cases presented a mix of *innovation triggers* of various kinds. Degradation of natural resources was among the most common triggers mentioned. Other common triggers included emergence of a local or global market opportunity, or introduction of a new technology or practice. Changes in policy were rarely mentioned.
- The relevant *time frame* for understanding the innovation process often exceeded ten years, and sometimes lasted for several decades.
- Many of the innovation processes involved several *interwoven dimensions*: technical (a new variety or technology), organizational (farmers acting collectively to acquire inputs or sell their produce) and institutional (new coordination mechanisms, new companies). These various dimensions did not usually emerge from the start: building on a specific entry point (typically a new technology), other dimensions emerged as the innovation process unfolded.

The JOLISAA inventory indicates that many African smallholders make efforts to counter degradation of the natural resources on which they depend and to link to markets to buy inputs and sell and transform their produce. New technologies are very important in shaping innovation, but organizational and, in some cases, institutional innovations also matter. By engaging with others, farmers receive much-needed support to pursue innovation while addressing the challenge of acquiring new capacities and skills to take advantage of these interactions. Many stakeholders with whom farmers and their organizations collaborate seem increasingly aware of the need for and advantages of such collaboration.

Source: Triomphe *et al.*, 2013.

political commitments and attitudes, and organizational dimensions.

Organizations dealing with rapid change must improve their capacity for continuous learning and innovation. Collective learning by organizations requires a combination of two elements: the ability to share knowledge; and the ability to make implicit knowledge explicit, so that an organization can digest it and transfer it across time (Ekboir *et al.*, 2009). This requirement implies reconsidering the role of M&E – which was traditionally designed to ensure better accountability – and moving towards a system that generates knowledge and facilitates learning. Given the methodological challenges of measuring impact, and the concern regarding capacity development, the focus is increasingly on measuring outcomes and identifying lessons for improving the innovation process (Klerkx and Gildemacher, 2012; Hall *et al.*, 2003).

However, measuring the capacity to innovate is in itself a challenge. Identifying appropriate indicators for tracking progress in capacity development and its process outcomes is not easy. As innovation programmes are based on complex processes at different levels and involving many stakeholders, there is need for mechanisms in which the performance of the entire learning, adaption and reflection process is regularly reviewed, and the activities, roles, relationships and effectiveness of different actors are evaluated.

The measurement and learning system needs to respond to the many different demands of the various stakeholders within the innovation system, and also of donors and development agencies where external funding is involved. Improving the design of the system requires both reducing its complexity by dividing it into discrete parts, with recognizable indicators attributable to specific interventions, and ensuring that these parts form a coherent whole. Essential elements include: (i) a knowledge and education domain – the research and education systems; (ii) a business and enterprise domain – value-chain stakeholders and family farmers; and (iii) bridging institutions – extension services, political channels and stakeholder platforms that link the two domains and facilitate the transfer of knowledge and information (Spielman and Birner, 2008). External influencing factors include linkages to other sectors of the economy (manufacturing and services); general science and technology policy; international actors, sources of knowledge and markets; and the political system.

Given the complexity of the challenge, governments need to take the lead in enabling and supporting M&E systems that facilitate access to and sharing of information and knowledge among and within these different elements of the innovation system, and are thus essential to a dynamic process of innovation benefiting family farms.

Key messages

- Capacity development for innovation should be based on a long-term strategy covering three interconnected dimensions: individual innovation capacity, organizational innovation capacity, and the creation of an enabling environment.
- At the individual level, greater investment is needed in human capital and education to support participants in the innovation system – family farmers, service providers, traders and processors, researchers, policy-makers, etc. – in developing their capacity to innovate. Special attention to youth and women is important. Sustained political support for investments in agricultural education and training is needed to develop a system of core institutions.
- At the organizational level, it is particularly important to support and facilitate the strengthening of producers' and other community-based organizations. Effective and inclusive producers' organizations can support innovation by their members, including by facilitating linkages with other actors in the innovation system – researchers, advisory service providers, value chains, etc. Particular emphasis must be put on women's inclusion in producers' organizations.
- At the system level, networks and linkages among different actors in the innovation system can facilitate the exchange of information and knowledge

and foster collaboration towards common goals. Useful mechanisms are innovation brokers – individuals or organizations that can bring different actors together – and innovation platforms, which provide a space for information sharing, negotiation, planning and action among different actors in an innovation system.

- The creation of an enabling environment for innovation is essential. This means that policies, incentives and governance mechanisms must improve the capacity of all actors in the innovation system to respond to change. Involving effective and representative producers' organizations in policy-making can ensure that public policies take into account the needs of family farms.
- There is need to learn from experiences and good practices in innovation and to develop capacity in measuring the impacts of different efforts and interventions to promote innovation capacity.

7. Conclusions: fostering innovation in family farming

Feeding the world in the next few decades will depend critically on the more than 500 million family farms that form the backbone of agriculture in most countries. These farmers are called on to produce much of the additional 60 percent of food[39] that the world's population will need by 2050. At the same time, family farms will have to play a leading role in the continuing fight against hunger and poverty and in preserving the natural environment against spreading degradation and advancing climate change.

Family farms are central to meeting some of the principal challenges that face the world in the twenty-first century. Their role derives in part from their sheer numbers – more than nine out of ten farms in the world are family farms – but it also stems from the huge potential of family farms to produce more food sustainably and to generate higher rural incomes.

The key to achieving this potential lies in innovation. For many small farms, innovation means moving away from growing food principally for their own consumption and going into commercial production. It means adopting new approaches, technologies and practices that not only increase production and efficiency, but also do so in full respect of natural processes and ecosystems.

However, if innovation is to take place on the farm, various changes must occur at other societal levels, including most obviously the public sector, where appropriate policies, funding and incentives must be in place, along with measures to encourage investment from the private sector. Government policies are often skewed in favour of large landowners and farms, and must be reoriented to foster innovation by smaller farmers.

[39] Compared with 2005/2007.

Innovation can occur only in the presence of well-functioning innovation systems whose various actors and components work together to bring beneficial change. Essential building blocks for innovation include well-run local government institutions, efficient agricultural advisory services, productive research and development centres, efficient producers' organizations, cooperatives and other community-based organizations, and – at the most basic level – an education system that fosters students' capacity to create and innovate.

Family farms already produce most of the world's food and occupy large tracts of the land, especially in developing countries. If they are to increase their contributions to food production and poverty reduction and act increasingly as stewards of the environment, they must be helped to face challenges in the best ways possible.

The changes required in family farming will involve more than the application of modern science, technology and marketing and management expertise. It will also be essential to farm more sustainably, in closer harmony with nature, and to re-evaluate traditional local knowledge and practices.

Innovation in agriculture cannot be viewed in isolation. Successful innovation must result in higher labour productivity among farming families to increase their incomes and reduce rural poverty. Labour productivity can also be enhanced by the availability of alternative and supplementary sources of employment and income for farming households. Appropriate measures for broader rural development that provide alternative livelihoods for farmers and other household members must be considered an integral part of promoting innovation in family farming.

Family farms are very diverse both among and within countries and communities, and they have different potentials and needs. This diversity calls for diversity in policy solutions; agricultural innovation systems and government support must be able to satisfy the different needs of different types of family farm. Some family farms are large commercial enterprises, which are likely to be already integrated into functioning agricultural innovation systems. Their main requirements are an enabling environment, adequate infrastructure and public agricultural research to ensure long-term production potential. They may also need appropriate incentives and regulations to motivate their adoption of sustainable practices that ensure the provision of essential environmental services (e.g. climate change mitigation, watershed protection, biodiversity conservation).

Some small and medium-sized family farms are already market-oriented and supply local, national or international markets; others have the potential to become commercial, given the right incentives, access to markets and support. These farms are less likely to be linked with agricultural innovation systems than are larger ones, but they may have significant potential for innovation. Helping this group of farmers to innovate can have a major impact on food security and can transform global agriculture. Special attention must be given to improving the innovation capacity of small and medium-sized farms and integrating them into innovation systems that are responsive to their needs. These efforts involve helping small and medium-sized farms overcome some of the constraints (financial limitations, high start-up costs, insecure property rights, etc.) that may prevent them from adopting improved practices. Farms also need agricultural research and inclusive advisory services that meet their needs and are suited to their specific circumstances. Farmers' organizations can play a central role in integrating small and medium-sized farmers into effective innovation systems.

Small subsistence family farms with limited commercial potential face similar constraints to innovation and have many of the same needs as small and medium-sized farms with commercial potential. However, most subsistence farms depend to a large extent on other, non-farming sources of income, and are unlikely to be able to emerge from poverty through agriculture alone. Reaching large numbers of these farmers and integrating them into effective agricultural innovation systems may be costly, hence the need to enhance social innovation and communication technologies to reduce costs. Collective action through farmers' organizations can help these farmers to innovate in agriculture and contribute to their livelihoods and food security. However, for most of them, escape from poverty requires efforts beyond agriculture and agricultural innovation, including overall rural development policies and effective social protection.

Governments need to develop their own strategies for different types of farmers, which also take into account social and equity dimensions. Governments have a clear responsibility for ensuring that rural areas and small family farms are not "forgotten", but the choice of policy instruments for supporting family farms will depend on national circumstances and governments' rural and overall development strategies and policy objectives.

It is important to remember that family farms are made up not only of crops and animals, but also of people. Within a family farm, different household members will relate to innovation systems in different ways and may have different needs. Capturing and considering these differences, particularly gender-based ones, is essential to making the innovation system more effective. Two groups of people are particularly important: women and young people. Women farmers generally face specific constraints to their productivity and capacity to innovate. Introducing a gender perspective into agricultural innovation systems will improve their effectiveness and enhance the productivity of family farms.

Young people are important because they often have an innate capacity to innovate that older household members may lack and because they represent the future of agriculture. Although youth can play a major role in ensuring that families connect to innovation systems, younger generations are increasingly leaving agriculture. In part, this is an expected feature of evolving economies. However, if youth come to see farming as a business, with real potential for innovation and profit, there may be positive effects on the prospects for growth and innovation in the sector.

Some of the key areas for promoting innovation in family farming for sustainable productivity growth are summarized in the following paragraphs.

Removing barriers and creating incentives for the adoption of technologies and practices for sustainable productivity growth. Farmers are ready to adopt new technologies and practices that they perceive as advantageous and are capable of implementing. However, several barriers make it difficult for farmers to adopt innovative processes, and women farmers face more of these barriers than do men.

Prohibitive impediments to sustainable productivity growth include the absence of marketing infrastructure, and insecure property and tenure rights. Another formidable barrier is the initial cost of adopting improved practices with long-term benefits, as this cost can be high and pay-off periods may be long. Long pay-off periods are particularly prohibitive when secure land tenure and access to financing and credit are lacking. Where innovative activities and practices generate public goods such as climate change mitigation, and incur significant costs, farmers will engage in them only if they are given appropriate compensation or incentives. As appropriate practices and technologies are often highly context-specific, the lack of solutions designed for local conditions can also be a serious impediment.

Local institutions, including producers' organizations, cooperatives and other community-based organizations, are central to farmers' ability to innovate. These institutions can play a key role in overcoming some of the barriers faced by small family farms in adopting improved practices. Where necessary, local institutions must be strengthened to facilitate smallholders' access to technical and management information, financing and markets. The effective functioning of local institutions, and their coordination with the public and private sectors and small farmers – men and women – are vital in helping small family farms adopt the innovative practices that will improve their own lives and their communities.

Investing in research and development. Investing in agricultural R&D is indispensable for sustaining and accelerating growth in agricultural productivity. The private sector can make an important contribution, and does so in many countries; however, because of the public good nature of much research, a strong public commitment to investing in R&D is needed. Such investments have high returns, but also generally long pay-off periods and uncertain benefits, especially for basic research. Long-term public commitment to continuous and stable funding of agricultural research is therefore fundamental. Flexible forms of shorter-term project or programme funding can contribute, but there needs to be a source of stable institutional financing to ensure long-term research capability.

Countries should carefully consider the best strategy for their specific needs and capacities. All countries need a certain level of domestic research capacity, but for those with limited financial resources and limited capacity to maintain strong national research programmes, the most effective strategy will be to tap into the results of international research and research by other countries and to focus on adapting these to their domestic circumstances. Other countries, with greater resources and less possibility of exploiting research by others, need to devote funds to more basic research. There is potential for South-South cooperation in agricultural research between countries with larger public-sector research institutes and smaller national agricultural research institutes in countries facing similar agro-ecological challenges. International partnerships and a careful division of labour between international research with broader applicability and national research geared to domestic needs are also needed.

There is need for research that is relevant to, and meets the specific needs of, family farms, especially smaller ones. Farmer-led innovation can make a major contribution, but needs to be supplemented by formal research. Linking scientific research to traditional knowledge can make research efforts more relevant and effective. Mechanisms and institutional arrangements must be in place to promote participatory research efforts involving family farmers and to ensure that family farms and their organizations are involved in setting research priorities and defining research agendas. It is critical that women farmers are also involved.

Developing agricultural extension and advisory services. Agricultural extension and advisory service are essential for promoting access to, and sharing of knowledge about, technologies and practices that support sustainable productivity growth among family farmers. However, many family farms lack regular access to extension. Modern extension services are characterized by the presence of a wide range of advisory services provided by a wide range of public, private and non-profit actors. Governments must facilitate the provision of advisory services by multiple actors, but have a responsibility to ensure that advisory services provided by the private sector and civil society are technically sound and socially and economically appropriate.

There is still a clear role for governments in providing agricultural advisory services. Such services can generate important public goods – increased productivity, improved sustainability, lower food prices, poverty reduction, etc. – which call for the involvement of the public sector. Providing services to small family farms, which are unlikely to be reached by commercial service providers, can be critical for poverty reduction and is clearly a government responsibility. However, governments have to consider the trade-offs between wide coverage of small or remote farms and the cost involved; in some cases, other instruments for rural poverty alleviation may be more cost-effective. Governments will need to make their own choices, based on national priorities. Government involvement is also necessary in the provision of advisory services for more sustainable agricultural practices, or for climate change adaptation and mitigation through reduced greenhouse gas emissions or increased carbon sequestration.

Ensuring the relevance and impact of rural advisory services means addressing the needs of different household members. Engaging women and ensuring that they have access to advisory services that respond to their specific needs and constraints is central. Participatory approaches such as Farmer Field Schools can be effective in engaging women and other household members in extension, but proactive measures may be needed to ensure women's participation.

Promoting capacity to innovate. The capacity to innovate should be promoted by developing individual and collective innovation capacity and creating an environment conducive to positive change. Some of the required interventions are specific to agriculture (e.g. agricultural training, promotion of producers' organizations); others are more generally beneficial (e.g. general education) and can help family farmers improve their farm productivity and increase and diversify their off-farm income.

At the individual level, skills and capacities must be upgraded by promoting education and training at all levels. Special attention must be given to girls, women, and youth in general. Education and training programmes that prepare young people to engage in commercial agriculture can determine future growth in the sector. An enabling environment for innovation includes good governance and economic policies, secure property rights, sound infrastructure and a conducive regulatory framework. Another key component is the building of networks and partnerships, in which different actors in the innovation system, including family farmers, can interact, share knowledge and experiences, and work towards shared goals.

An essential element is the building and strengthening of producers' organizations. Strong, effective and inclusive producers' organizations can have a major impact on the capacity of family farms to innovate. They can facilitate farmers' access to markets, giving them incentives to innovate; serve as a vehicle for closer cooperation with national research institutes; provide extension and advisory services to their members, and serve as intermediaries between individual family farms and other rural advisory service providers; and ensure that family farms have a voice in policy debates and can influence national priorities for innovation. Effective engagement of both women and men should be pursued, while measures should be taken to avoid elite capture by larger and more influential farmers.

Key messages of the report

The State of Food and Agriculture 2014: Promoting innovation in family farming offers the following key messages:

- **Family farms are part of the solution for achieving food security and sustainable rural development; the world's food security and environmental sustainability depend on the more than 500 million family farms that form the backbone of agriculture in most countries.** Family farms represent more than nine out of ten farms in the world and can serve as a catalyst for sustained rural development. They are the stewards of the world's agricultural resources and the source of more than 80 percent of the world's food supply, but many of them are poor and food-insecure themselves. Innovation in family farming is urgently needed to lift farmers out of poverty and help the world achieve food security and sustainable agriculture.
- **Family farms are an extremely diverse group, and innovation systems must take this diversity into account.** Innovation strategies for all family farms must consider their agro-ecological and socio-economic conditions and government policy objectives for the sector. Public efforts to promote agricultural innovation for small and medium-sized family farms should ensure that agricultural research, advisory services, market institutions and infrastructure are inclusive. Applied agricultural research for crops, livestock species and management practices of importance to these farms are public goods and should be a priority. A supportive environment for producers' and other community-based organizations can help promote innovation, through which small and medium-sized family farms could transform world agriculture.
- **The challenges facing agriculture and the institutional environment for agricultural innovation are far more complex than ever before; the world must create an innovation system that embraces this complexity.** Agricultural innovation strategies must now focus not just on increasing yields but also on a more complex set of objectives, including preserving natural resources and raising rural incomes. They must also take into account today's complex policy and institutional environment for agriculture and the more pluralistic set of actors engaged in decision-making. An *innovation system* that facilitates and coordinates the activities of all stakeholders is essential.
- **Public investment in agricultural R&D and extension and advisory services should be increased and refocused to emphasize sustainable intensification and closing yield and labour productivity gaps.** Agricultural research and advisory services generate public goods – productivity, improved sustainability, lower food prices, poverty reduction, etc. – calling for strong government involvement. R&D should focus on sustainable intensification, continuing to expand the production frontier but in sustainable ways, working at the system level and incorporating traditional knowledge. Extension and advisory services should focus on closing yield gaps and raising the labour productivity of small and medium-sized farmers. Partnering with producers' organizations can help ensure that R&D and extension services are inclusive and responsive to farmers' needs.
- **All family farmers need an enabling environment for innovation, including good governance, stable macroeconomic conditions, transparent legal and regulatory regimes, secure property rights, risk management tools and market infrastructure.** Improved access to local or wider markets for inputs and outputs, including through government procurement from family farmers, can provide strong incentives for innovation, but farmers in remote areas and marginalized groups often face severe barriers. In addition, sustainable agricultural practices often have high start-up costs and long pay-off periods and farmers may need appropriate incentives to provide important environmental services.

Effective local institutions, including farmers' organizations, combined with social protection programmes, can help overcome these barriers.

- **Capacity to innovate in family farming must be promoted at multiple levels.** Individual innovation capacity must be developed through investment in education and training. Incentives are needed for the creation of networks and linkages that enable different actors in the innovation system – farmers, researchers, advisory service providers, value chain participants, etc. – to share information and work towards common objectives.
- **Effective and inclusive producers' organizations can support innovation by their members.** Producers' organizations can assist their members in accessing markets and linking with other actors in the innovation system. They can also help family farms have a voice in policy-making.

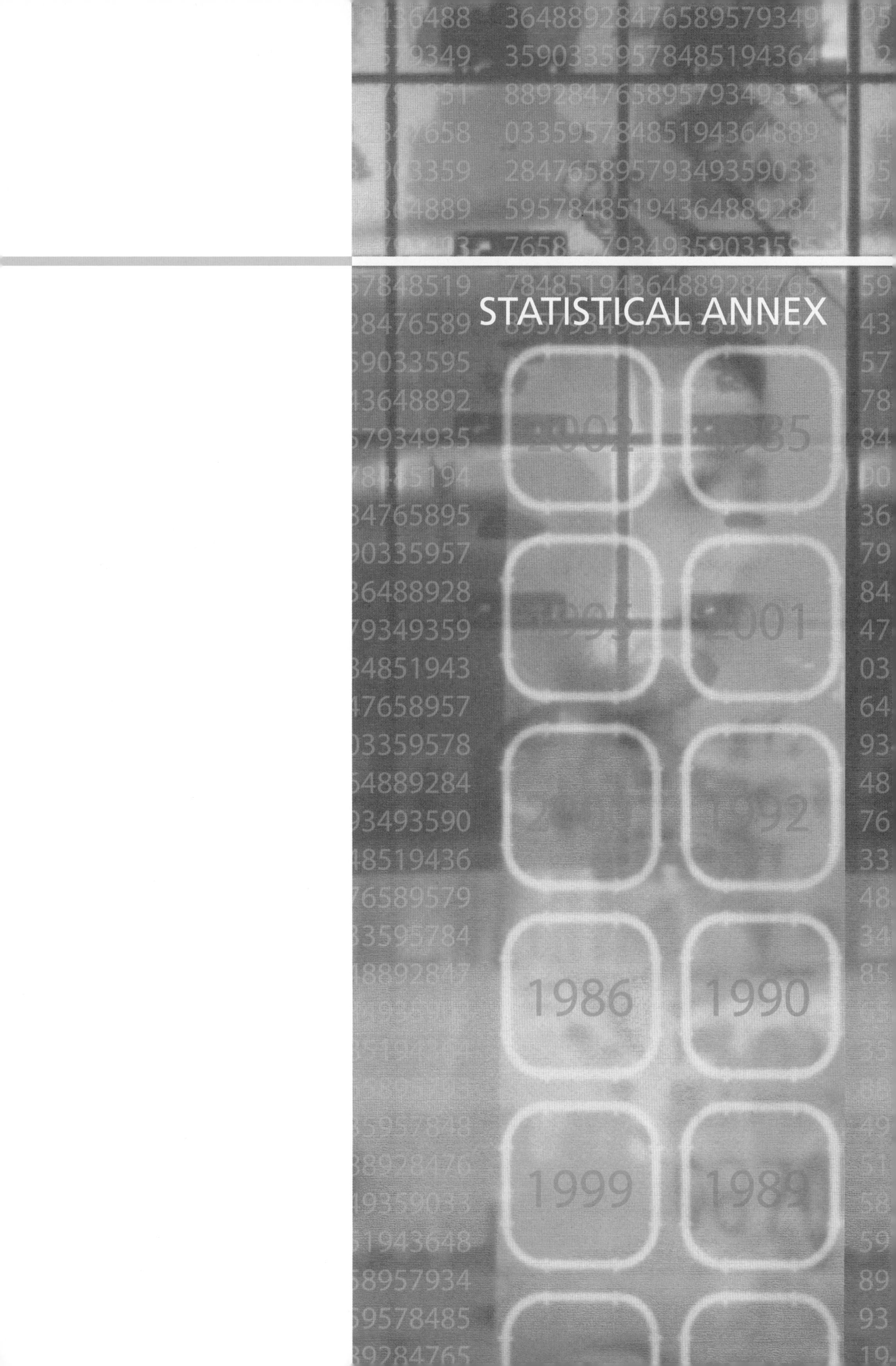

STATISTICAL ANNEX

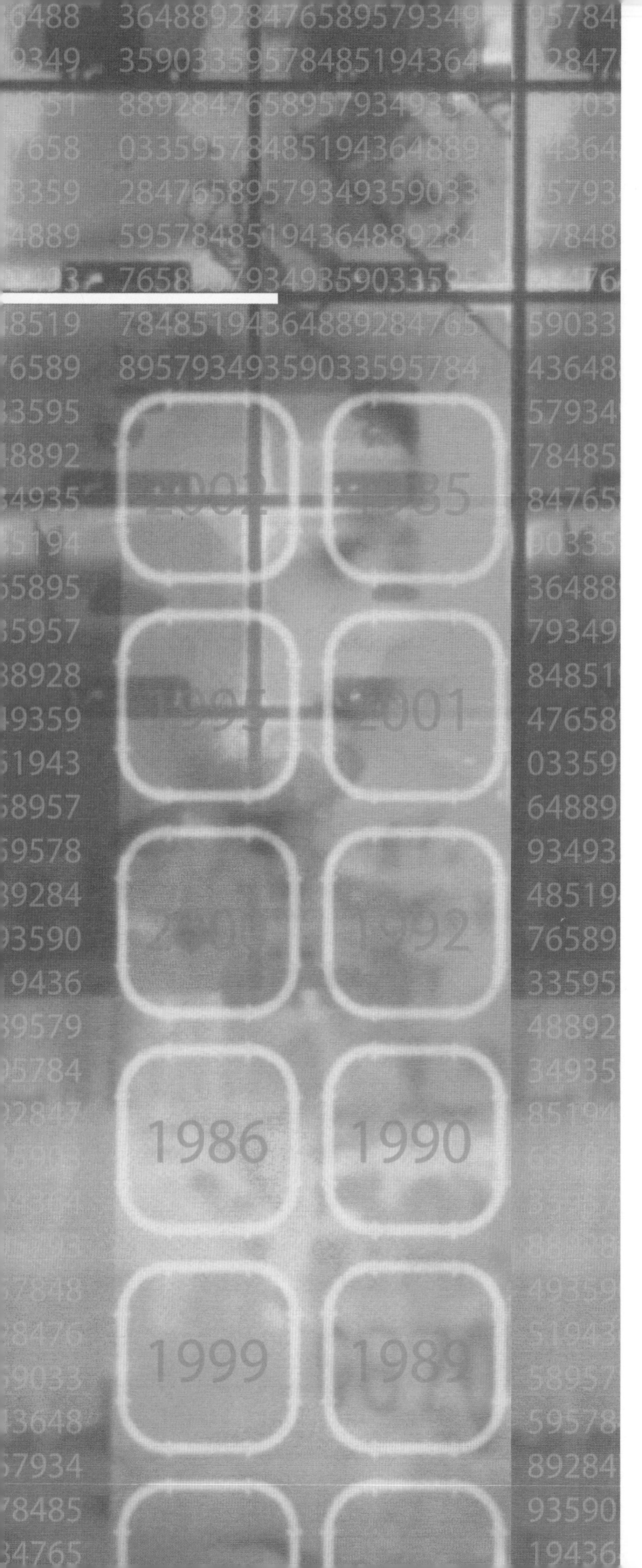
2001
1992
1986
1990
1999
1989

Notes on the annex tables

Key

The following conventions are used in the tables:

.. = data not available
0 or 0.0 = nil or negligible
blank cell = not applicable

Numbers presented in the tables may differ from the original data sources because of rounding or data processing. To separate decimals from whole numbers a full point (.) is used.

Technical notes

Table A1. Number of agricultural holdings and size of agricultural area

Sources: Data on the number of holdings was compiled by the authors using FAO (2013a), FAO (2001) and other sources from the FAO Programme for the World Census of Agriculture. Full documentation is provided below. Data on agricultural area are from FAO (2014).

For Table A1 the world total agricultural area equals the sum of regional subtotals; it is slightly larger than the sum of income grouping subtotals, since the regional groupings include some countries and territories that are not included in the income classification.

Agricultural holdings

Agricultural holdings reported by agricultural censuses include crop and livestock production only; holdings engaged in forestry or fisheries are only included if they also are engaged in crop and livestock production. An agricultural holding is an economic unit of agricultural production under single management comprising all livestock kept and all land used wholly or partly for agricultural production purposes, without regard to title, legal form, or size. Single management may be exercised by an individual or a household, jointly by two or more individuals or household, by a clan or tribe, or by a juridical person such as a corporation or a government agency. The holding's land may consist of one or more parcels, located in one or more separate areas or in one or more territorial or administrative divisions, provided the parcels share the same production means utilized by the holding, such as labour, farm building, machinery or draught animals. For a limited number of countries, the number of holdings was not available and the number of agricultural households is therefore presented in Table A1.

Agricultural area
Agricultural area reported by FAOSTAT is the sum of areas under: (a) arable land, (b) permanent crops, and (c) permanent meadows and pastures. "Arable land" is land under temporary agricultural crops (multiple-cropped areas are counted only once), temporary meadows for mowing or pasture, land under market and kitchen gardens and land temporarily fallow (less than five years). The abandoned land resulting from shifting cultivation is not included in this category. "Permanent crops" refers to land cultivated with long-term crops which do not have to be replanted for several years (such as cocoa and coffee); land under trees and shrubs producing flowers; and nurseries (except those for forest trees, which should be classified under "forest"). "Permanent meadows and pastures" refers to land used permanently (five years or more) to grow herbaceous forage crops, either cultivated or growing wild (wild prairie or grazing land).

Table A2. Shares of agricultural holdings and agricultural area, by land size class

Source: Authors' compilation using most recent data from the FAO Programme for the World Census of Agriculture 1990 or 2000 rounds; as shown in FAO (2001) and FAO (2013a).

Table A2 covers the 106 countries for which data on the number of holdings by land size class are available, although data on agricultural area by land size class are not available for all countries. Figure 2 covers all 106 countries, aggregated at the world level, and includes estimates for agricultural area by land class size for those countries for which data are not available. These estimates are not reported in Table A2. See Lowder, Skoet and Singh (2014) for details. Figure 3 covers only those countries in Table A2 for which data on both the number of holdings and agricultural area by land size are available and for which the World Bank assigned an income classification in 2011 (see World Bank [2012]).

The land size classes reported in Table A2 and Figures 2 and 3 are those most commonly used in national agricultural census reports, and adjustments have been made for some countries which report different land size classes. For example, some countries do not report a land size class of less than 1 hectare; rather they use a larger minimum cut-off point. In such cases, all farms below the minimum cut-off point are shown in the smallest class size reported by the country although some farms may be smaller than 1 hectare. Similarly, some countries do not report a land size class of greater than 50 hectares. In such cases, all farms larger than the national cut-off point are included in the largest land-size class shown for the country even though some farms may be larger than 50 hectares.

Holdings
Holdings refers to share of *agricultural holdings* included in each land size class; for definition see notes to Table A1.

Area
Refers to the share of *area of holdings* in each land size class. For some countries that do not report area of holding, the table presents a partial measure, such as the agricultural area, agricultural land, cropland, utilized agricultural area or other.

Area of holding provides the most comprehensive measure of the size of the holding. It refers to all land managed and operated by an agricultural holding, without regard to the right to access the land. It includes the land owned by the holder plus the land rented-in plus the land operated under other forms of tenure. This should not be confused with "agricultural areas", which is a subcategory of operated area of the holdings.

Agricultural area or *agricultural land* consists of cropland and permanent meadows and pastures.

Cropland consists of arable land plus land under permanent crops.

Utilized agricultural area includes arable land, kitchen gardens, permanent meadows and pastures, and permanent crops.

For details, refer to the original source FAO (2013a) and FAO (2001), as well as FAO (2005).

Table A3. Average level and rate of change in agricultural labour productivity, 1961–2012

Source: Authors' calculations using FAO (2014) and FAO (2008a). Table A3 includes only those countries for which the World Bank assigned an income classification in 2011 (see World Bank [2012]).

Agricultural labour productivity

The value of agricultural production divided by the population economically active in agriculture. The value of agricultural production is the net production value measured in constant 2004–06 international dollars. The value of net production is compiled by multiplying gross crop and livestock production in physical terms by output prices at farm gate and subtracting intermediate uses within the agricultural sector (such as seed and feed). The economically active population in agriculture (agricultural labour force or agricultural workers) is that part of the economically active population engaged in or seeking work in agriculture, hunting, fishing or forestry.

Rate of change in agricultural labour productivity

The average annual rate of change is estimated using the OLS regression method; that is, the natural logarithm of the value of agricultural production is regressed on a variable for time and a constant term for all available observations in the decade.

Regional and income groupings

Countries are listed in alphabetical order according to the income and regional groupings established by the World Bank country classification system in July 2011; see World Bank (2012a) for a description. The World Bank does not provide an income classification for the following seven geographical entities: the Cook Islands, French Guiana, Guadeloupe, Martinique, Nauru, Niue and Réunion. Those entities are therefore not included in the totals or averages by income grouping, but they are included in regional totals or averages.

Country notes

Data for China exclude data for Hong Kong Special Administrative Region of China and Macao Special Administrative Region of China.

Whenever possible, data from 1992 or 1995 onwards are shown for Armenia, Azerbaijan, Belarus, Estonia, Georgia, Kazakhstan, Kyrgyzstan, Latvia, Lithuania, Moldova, Russian Federation, Tajikistan, Turkmenistan, Ukraine and Uzbekistan. Available data for years prior to 1992 are shown for the Union of Soviet Socialist Republics ("USSR" in the table listings).

Data for years prior to 1992 are provided for the former Yugoslavia ("Yugoslavia SFR" in the table listings).

Observations for the years following 1992 are provided for the individual countries formed from the former Yugoslavia; these are Bosnia and Herzegovina, Croatia, the former Yugoslav Republic of Macedonia, and Slovenia, as well as Serbia and Montenegro. Observations are provided separately for Serbia and for Montenegro after the year 2006.

Data are shown when possible for the individual countries formed from the former Czechoslovakia – the Czech Republic and Slovakia. Data for years prior to 1993 are shown under Czechoslovakia.

Data are shown for Eritrea and Ethiopia separately, if possible; in most cases before 1992 data on Eritrea and Ethiopia are aggregated and presented as Ethiopia PDR.

Data for Yemen refer to that country from 1990 onward; data for previous years refer to aggregated data of the former People's Democratic Republic of Yemen and the former Yemen Arab Republic.

Separate observations are shown for Belgium and Luxembourg whenever possible.

Sources for Table A1

1. **FAO.** 2013a. *2000 World Census of Agriculture. Analysis and international comparison of the results (1996–2005).* FAO Statistical Development Series 13. Rome.

2. **Government of China.** 2009. *Abstract of the Second National Agricultural Census in China 2006.* Beijing, National Bureau of Statistics of China.

3. **Government of Fiji.** 2009. *Fiji National Agricultural Census 2009.* Suva, Fiji, Department of Agriculture.

4. **Government of Lao People's Democratic Republic.** 2012. *Lao Census of Agriculture 2010/11. Highlights. Summary census report.* Vientiane, Ministry of Agriculture and Forestry.

5. **Government of Myanmar.** 2013. *Report on Myanmar Census of Agriculture (MCA) 2010.* Myanmar, Ministry of Agriculture and Irrigation.

6. **Government of Niue.** 2009. *Agricultural Census of Niue 2009.* Niue, Department of Agriculture, Forestry and Fisheries.

7. **FAO.** 2001. *Supplement to the report on the 1990 World Census of Agriculture. International comparison and primary results by country (1986–1995).* FAO Statistical Development Series 9a. Rome.

8. **Government of Samoa.** 2012. *Analytical report of the 2009 Census of Agriculture.* Apia, Samoa Bureau of Statistics.

9. **Government of Vanuatu.** 1993. *Vanuatu Agricultural Census 1993. Main results.* Port Vila, Vanuatu National Statistics Office.
10. **Government of Albania.** 2012. *Preliminary results of agriculture census, 2012.* Tirana, Instituti i Statistikave.
11. **European Union.** 2012. *Agriculture, fishery and forestry statistics. Main results 2010–11. Eurostat Pocketbooks.* Luxembourg.
12. **Government of Montenegro.** 2011. *Popis poljoprivrede 2010. Struktura poljoprivrednih gazdinstava. Znamo šta imamo.* Podgorica, Statistical Office of Montenegro.
13. **Government of Republic of Moldova.** 2011. *Recensămîntul general agricol 2011. Rezultate preliminare.* Chişinău, Biroul Naţional de Statistică al Republicii.
14. **Government of the Russian Federation.** 2008. *2006 All-Russia Census of Agriculture: Russian Federation summary and country-level data.* Federal State Statistics Service. Moscow, Statistics of Russia Information and Publishing Center.
15. **Government of Republic of Macedonia.** 2007. *Census of Agriculture, 2007. Basic statistical data on individual agricultural holdings and business entities in the Republic of Macedonia, by regions. Book I.* Skopje, State Statistical Office of the Republic of Macedonia.
16. **Government of Argentina.** 2009. *Censo Nacional Agropecuario 2008–CNA '08. Resultados provisorios.* Buenos Aires, Instituto Nacional de Estadística y Censos.
17. **Government of Brazil.** 2009. *Censo Agropecuário 2006.* Rio de Janeiro, Instituto Brasileiro de Geografia e Estatística (IBGE).
18. **Government of Chile.** 2007. *VII Censo Nacional Agropecuario y Forestal.* Santiago, Instituto Nacional de Estadísticas.
19. **Government of El Salvador.** 2009. *IV Censo Agropecuario 2007–2008. Resultados nacionales.* San Salvador, Ministerio de Economía.
20. **Government of France.** 2011. *Agreste: la statistique agricole. Numéro 02, Novembre 2011. Premières tendances, recensement agricole 2010 Guyane.* Press report. Cayenne, French Guiana, Ministère de l'Agriculture, de l'Alimentation, de la Pêche, de la Ruralité et de l'Aménagement du Territoire.
21. **Government of France.** 2011. *Agreste: la statistique agricole. Numéro 10, septembre 2011. Premières tendances, recensement agricole 2010 Guadeloupe.* Basse-terre, Ministère de l'Agriculture, de l'Alimentation, de la Pêche, de la Ruralité et de l'Aménagement du Territoire.
22. **Government of Haiti.** 2012. *Synthèse nationale des résultats du Recensement Général de L'agriculture (RGA) 2008–2009.* Port-au-Prince, Ministère de l'Agriculture des Ressources Naturelles et du Développement Rural.
23. **Government of Jamaica.** 2007. *Census of Agriculture 2007. Preliminary report.* Kingston, The Statistical Institute of Jamaica.
24. **Government of France.** 2011. *Agreste: la statistique agricole. Numéro 7, septembre 2011. Premières tendances, recensement agricole 2010 Martinique.* Press report. Fort-de-France, Ministère de l'Agriculture, de l'Alimentation, de la Pêche, de la Ruralité et de l'Aménagement du Territoire.
25. **Government of Mexico.** 2009. *VIII Censo Agrícola, Ganadero y Forestal 2007.* Aguascalientes, Instituto Nacional de Estadística y Geografía (INEGI).

26. **Government of Nicaragua.** 2012. *IV Censo Nacional Agropecuario (CENAGRO 2011). Informe final.* Managua, Instituto Nacional de Información de Desarrollo (INIDE).
27. **Government of Panama.** 2012.*VII Censo Nacional Agropecuario, 2011.* Vol. I, *Resultados finales basicos.* Panama City, Instituto Nacional de Estadistica y Censo.
28. **Government of Paraguay.** 2009. *Censo Agropecuario Nacional 2008.* Vol. I. San Lorenzo, Ministerio de Agricultura y Ganadería.
29. **Government of Peru.** 2012. *IV Censo Nacional Agropecuario 2012. Resultados preliminares.* Lima, Instituto Nacional de Estadística e Informática (INEI).
30. **Government of Saint Lucia.** 2007. *St. Lucia Census of Agriculture. Final report 2007.* Saint Lucia, Ministry of Agriculture, Forestry and Fisheries.
31. **Government of Uruguay.** 2012. *Presentación de datos preliminares del Censo General Agropecuario 2011.* Montevideo, Estadísticas Agropecuarias (DIEA), Ministerio de Ganadería Agricultura y Pesca.
32. **Government of Venezuela.** 2008. *VII Censo Agrícola Nacional (Mayo 2007/Abril 2008).* Caracas, Ministerio del Poder Popular para la Agricultura y Tierras.
33. **Government of Jordan.** 2007. *Agricultural Census 2007. Provisional tables.* Amman, Department of Statistics.
34. **Government of Bangladesh.** 2010. *Census of Agriculture 2008. Structure of agricultural holdings and livestock population. Vol. 1.* Dhaka, Bangladesh Bureau of Statistics.
35. **Kingdom of Bhutan.** 2010. *Renewable Natural Resources (RNR) Census 2009. Vol. 1.* Thimpu, Ministry of Agriculture and Forests.
36. **Government of India.** 2012. *Agriculture Census 2010–11 Phase 1. All India report on number and area of operational holdings (Provisional).* New Delhi, Ministry of Agriculture.
37. **Government of Malawi.** 2010. *National Census of Agriculture and Livestock 2006–7. Main report.* Zomba, National Statistical Office.
38. **Government of France.** 2011. *Agreste: la statistique agricole. Mémento 2011 La Réunion. Numéro 75, février 2012.* Saint Denis, Direction de l'Alimentation, de l'Agriculture et de la Forêt de La Réunion.
39. **Government of Rwanda.** 2010. *National Agricultural Survey 2008 (NAS 2008).* Kigali, National Institute of Statistics of Rwanda.
40. **Government of Belgium.** 2011. *Résultats définitifs de L'enquête Agricole de mai 2010.* Communiqué de presse 11 mars 2011. Belgium. SPF Économie PME, Classes Moyennes et Énergie (ECONOMIE), Direction Générale Statistique et Information Économique.
41. **Government of Cyprus.** 2012. *Agricultural statistics 2009–2010.* Series II, Report no. 41. Nicosia, Statistical Service.
42. **Government of Czech Republic.** 2011. *Agrocensus 2010. Farm structure survey and survey on agricultural production methods. Environment, Agriculture. Volume 2011.* Prague, Agricultural, Forestry and Environmental Statistics Department.
43. **Government of Finland.** 2013. *Maatalouslaskenta 2010. Agricultural Census. Agricultural and horticultural enterprises, labour force and diversified farming.* Helsinki, Information Centre of the Ministry of Agriculture and Forestry (TiKe).
44. **U.S. Department of Agriculture (USDA).** 2009. *2007 Census of Agriculture. Guam. Island data. Geographic Area Series, Vol. 1, Part 53.* National Agricultural Statistics Service (NASS).

45. **Government of Malta.** 2012. *Census of Agriculture 2010.* Valletta, National Statistics Office.
46. **USDA.** 2009. *2007 Census of Agriculture. Northern Mariana Islands. Commonwealth and Island Data. Geographic Area Series, Vol. 1, Part 56.* Washington, DC, National Agricultural Statistics Service (NASS).
47. **Government of Slovenia.** 2012. *The 2010 Agricultural Census. Every farm counts!* Brochure. Ljubljana, Statistical Office of the Republic of Slovenia.
48. **USDA.** 2009. *2007 Census of Agriculture. United States. Summary and state data.* Geographic Area Series, Vol. 1, Part 51. Washington, DC, National Agricultural Statistics Service (NASS).
49. **USDA.** 2009. *2007 Census of Agriculture. Virgin Islands of the United States. Territory and island data.* Geographic Area Series, Vol. 1, Part 54. Washington, DC, National Agricultural Statistics Service (NASS).

TABLE A1
Number of agricultural holdings and size of agricultural area

	Number of holdings (Thousands)	Census year/ round	Source	Agricultural area (Thousand ha)					
				1961	1971	1981	1991	2001	2011
LOW-INCOME COUNTRIES	71 522			544 378	555 942	561 262	572 059	592 129	619 851
LOWER-MIDDLE-INCOME COUNTRIES	208 148			776 999	792 253	795 124	828 476	966 626	837 233
UPPER-MIDDLE-INCOME COUNTRIES	268 035			1 834 035	1 930 608	2 021 725	2 141 242	2 054 897	2 063 966
HIGH-INCOME COUNTRIES	21 867			1 297 955	1 294 798	1 282 444	1 290 691	1 315 429	1 246 991
WORLD	569 600			4 453 535	4 573 782	4 660 737	4 832 652	4 929 245	4 768 186
LOW- AND MIDDLE-INCOME COUNTRIES	547 706			3 155 412	3 278 803	3 378 111	3 541 777	3 613 651	3 521 049
East Asia and the Pacific	253 837			571 515	611 593	657 205	746 607	770 859	764 584
American Samoa	7	2003	1	3	3	3	3	5	5
Cambodia	..	..		3 518	2 450	2 650	4 510	4 890	5 655
China	200 555	2006	2	343 248	380 165	433 818	510 896	524 099	519 148
Cook Islands	2	2000	1	6	6	6	6	6	3
Democratic People's Republic of Korea	..	..		2 380	2 380	2 515	2 530	2 550	2 555
Fiji	65	2009	3	227	221	300	424	428	428
Indonesia	24 869	2003	1	38 600	38 350	37 950	41 524	46 300	54 500
Kiribati	..	..		39	38	38	39	34	34
Lao People's Democratic Republic	783	2010–11	4	1 550	1 482	1 609	1 662	1 839	2 378
Malaysia	526	2005	1	4 200	4 721	5 121	7 475	7 870	7 870
Marshall Islands	..	..		..	..	..	12	12	13
Micronesia (Federated States of)	..	..		..	..	..	23	23	22
Mongolia	250	2000	1	140 683	140 683	124 519	126 130	129 704	113 507
Myanmar	5 426	2010	5	10 430	10 805	10 421	10 416	10 939	12 558
Nauru	..	..		0	0	0	0	0	0
Niue	0	2009	6	3	4	5	5	5	5
Palau	0	1990	7	..	..	..	5	5	5
Papua New Guinea	..	..		495	669	778	882	1 010	1 190
Philippines	4 823	2002	1	7 713	8 279	10 670	11 157	11 134	12 100
Samoa	16	2009	8	56	64	77	54	48	35
Solomon Islands	..	..		55	55	59	69	77	91
Thailand	5 793	2003	1	11 653	14 399	19 341	21 516	19 828	21 060
Timor-Leste	..	..		230	243	282	330	362	360
Tonga	11	2001	1	27	32	34	32	30	31
Tuvalu	..	..		2	2	2	2	2	2

TABLE A1 *(cont.)*

	Number of holdings *(Thousands)*	Census year/ round	Source	Agricultural area *(Thousand ha)*					
				1961	1971	1981	1991	2001	2011
Vanuatu	22	1993	9	105	120	131	154	177	187
Viet Nam	10 690	2001	1	6 292	6 422	6 876	6 751	9 483	10 842
Europe and Central Asia	**37 342**			**614 775**	**622 578**	**628 637**	**631 544**	**637 138**	**632 694**
Albania	324	2012	10	1 232	1 200	1 116	1 127	1 139	1 201
Armenia	..	..						1 328	1 711
Azerbaijan	1 287	2004–05	1					4 746	4 769
Belarus	..	..						9 128	8 875
Bosnia and Herzegovina	..	..						2 126	2 151
Bulgaria	370	2010	11	5 673	6 009	6 179	6 161	5 498	5 088
Georgia	730	2003–04	1					3 003	2 469
Kazakhstan	..	..						207 269	209 115
Kyrgyzstan	1 131	2002	1					10 776	10 609
Latvia	180	2001	1					1 581	1 816
Lithuania	611	2003	1					2 896	2 806
Montenegro	49	2010	12						512
Republic of Moldova	902	2011	13					2 539	2 459
Romania	4 485	2002	1	14 601	14 935	14 948	14 798	14 798	13 982
Russian Federation	23 224	2006	14					216 861	215 250
Serbia	779	2002	1						5 061
Serbia and Montenegro								5 592	
Tajikistan	..	..						4 573	4 855
The former Yugoslav Republic of Macedonia	193	2007	15					1 242	1 118
Turkey	3 077	2001	1	36 517	38 314	38 613	40 067	40 968	38 247
Turkmenistan	..	..						32 360	32 660
Ukraine	..	..						41 385	41 281
USSR				541 800	547 600	553 500	555 420		
Uzbekistan	..	..						27 330	26 660
Yugoslav SFR				14 952	14 520	14 281	13 971		
Latin America and the Caribbean	**21 022**			**559 454**	**612 767**	**652 864**	**688 275**	**708 496**	**739 589**
Antigua	5	1980	7	..	..	..	..	..	..
Antigua and Barbuda	..	..		10	11	7	9	9	9
Argentina	277	2008	16	137 829	129 154	127 894	127 660	128 606	147 548
Belize	11	1980	7	79	83	97	130	149	157
Bolivia (Plurinational State of)	..	..		30 042	30 734	34 099	35 796	37 006	37 055

TABLE A1 *(cont.)*

	Number of holdings *(Thousands)*	Census year/ round	Source	Agricultural area *(Thousand ha)*					
				1961	1971	1981	1991	2001	2011
Brazil	5 175	2006	17	150 531	199 632	225 824	244 941	263 465	275 030
Chile	301	2007	18	13 386	15 350	16 750	15 789	15 150	15 789
Colombia	2 022	2001	1	39 970	45 054	45 308	44 884	41 745	43 786
Costa Rica	82	1970	7	1 395	1 887	2 599	2 238	1 833	1 880
Cuba	..	..		3 550	5 073	5 938	6 755	6 656	6 570
Dominica	9	1995	7	17	19	19	18	22	26
Dominican Republic	305	1970	7	2 190	2 344	2 625	2 570	2 515	2 447
Ecuador	843	1999–2000	1	4 710	4 915	6 759	7 914	7 785	7 346
El Salvador	397	2008	19	1 252	1 278	1 370	1 428	1 550	1 532
French Guiana	6	2010	20	6	7	9	21	23	23
Grenada	18	1995	7	22	22	16	12	13	11
Guadeloupe	8	2010	21	58	63	59	53	48	42
Guatemala	831	2003	1	2 646	2 767	3 067	4 285	4 495	4 395
Guyana	..	..		1 359	1 371	1 715	1 734	1 708	1 677
Haiti	1 019	2008	22	1 660	1 710	1 600	1 596	1 670	1 770
Honduras	326	1993	7	2 980	3 045	3 264	3 342	2 936	3 220
Jamaica	229	2007	23	533	507	497	476	479	449
Martinique	3	2010	24	34	38	38	36	33	27
Mexico	5 549	2007	25	98 244	97 779	99 249	104 500	105 400	103 166
Nicaragua	269	2011	26	3 430	3 605	3 827	4 060	5 144	5 146
Panama	249	2011	27	1 624	1 713	1 882	2 134	2 243	2 267
Paraguay	290	2008	28	10 411	11 518	13 457	17 195	20 200	20 990
Peru	2 293	2012	29	16 956	17 922	18 704	21 896	21 150	21 500
Saint Lucia	9	2007	30	17	20	20	20	14	11
Saint Vincent and the Grenadines	7	2000	1	10	11	12	12	10	10
Suriname	22	1980	7	41	52	73	89	86	82
Uruguay	45	2011	31	15 230	15 057	15 046	14 825	14 955	14 378
Venezuela (Bolivarian Republic of)	424	2007–08	32	19 232	20 026	21 040	21 857	21 398	21 250
Middle East and North Africa	**14 927**			**200 889**	**206 641**	**203 359**	**209 384**	**212 067**	**198 895**
Algeria	1 024	2001	1	45 471	45 433	39 171	38 622	40 109	41 383
Djibouti	1	1995	7	1 301	1 301	1 301	1 336	1 681	1 702
Egypt	4 542	1999–2000	1	2 568	2 852	2 468	2 643	3 338	3 665
Iran (Islamic Republic of)	4 332	2003	1	59 271	60 154	58 280	62 997	63 823	48 957
Iraq	591	1970	7	8 800	8 999	9 439	9 630	8 490	8 210
Jordan	80	2007	33	1 084	1 105	1 118	1 010	1 022	1 003

TABLE A1 *(cont.)*

	Number of holdings *(Thousands)*	Census year/ round	Source	Agricultural area *(Thousand ha)*					
				1961	1971	1981	1991	2001	2011
Lebanon	195	1998	1	562	630	598	606	598	638
Libya	176	1987	7	11 170	13 235	15 185	15 460	15 450	15 585
Morocco	1 496	1996	1	23 370	26 812	29 090	30 355	30 370	30 104
Occupied Palestinian Territory	..	..		366	368	379	372	369	261
Syrian Arab Republic	486	1980	7	14 941	13 458	14 115	13 512	13 723	13 864
Tunisia	516	2004	1	8 648	8 868	8 750	9 210	9 499	10 072
Yemen	1 488	2002	1	23 337	23 426	23 465	23 631	23 595	23 452
South Asia	**169 295**			**249 588**	**256 117**	**260 818**	**262 454**	**261 843**	**260 793**
Afghanistan	3 045	2002	1	37 700	38 036	38 053	38 030	37 753	37 910
Bangladesh	15 183	2008	34	9 480	9 695	9 981	10 320	9 403	9 128
Bhutan	62	2009	35	361	382	413	504	535	520
India	137 757	2011	36	174 907	177 700	180 459	181 140	180 370	179 799
Maldives	..	..		5	6	7	8	10	7
Nepal	3 364	2002	1	3 531	3 680	4 216	4 150	4 261	4 259
Pakistan	6 620	2000	1	21 881	24 279	25 340	25 960	27 160	26 550
Sri Lanka	3 265	2002	1	1 723	2 339	2 349	2 342	2 351	2 620
Sub-Saharan Africa	**51 309**			**959 359**	**969 287**	**975 410**	**1 003 697**	**1 023 413**	**924 641**
Angola	1 067	1970	7	57 170	57 400	57 400	57 450	57 300	58 390
Benin	408	1990	7	1 442	1 777	2 057	2 280	3 265	3 430
Botswana	51	2004	1	26 000	26 001	26 004	25 901	25 801	25 861
Burkina Faso	887	1993	7	8 139	8 220	8 835	9 550	10 660	11 765
Burundi	..	..		1 575	1 899	2 150	2 125	2 307	2 220
Cape Verde	45	2004	1	65	65	65	68	73	75
Cameroon	926	1970	7	7 510	8 028	8 960	9 150	9 160	9 600
Central African Republic	304	1980	7	4 738	4 840	4 945	5 008	5 149	5 080
Chad	366	1970	7	47 900	47 900	48 150	48 350	48 930	49 932
Comoros	52	2004	1	95	105	110	133	147	155
Congo	143	1980	7	10 540	10 548	10 528	10 523	10 540	10 560
Côte d'Ivoire	1 118	2001	1	15 680	16 300	17 370	18 950	19 600	20 500
Democratic Republic of the Congo	4 480	1990	7	25 050	25 400	25 750	25 980	25 550	25 755
Eritrea	..	..						7 532	7 592
Ethiopia	10 759	2001–02	1					31 409	35 683
Ethiopia PDR				57 836	59 340	58 860	56 158		
Gabon	71	1970	7	5 195	5 200	5 152	5 157	5 160	5 160
Gambia	69	2001–02	1	524	537	585	592	560	615

TABLE A1 *(cont.)*

	Number of holdings *(Thousands)*	Census year/ round	Source	Agricultural area *(Thousand ha)*					
				1961	1971	1981	1991	2001	2011
Ghana	1 850	1980	7	11 700	11 700	12 000	12 720	14 510	15 900
Guinea	840	2000–01	1	14 620	14 405	14 197	14 049	13 540	14 240
Guinea-Bissau	84	1988	7	1 358	1 368	1 390	1 447	1 628	1 630
Kenya	2 750	1980	7	25 200	25 250	25 580	26 877	26 839	27 450
Lesotho	338	1999-2000	1	2 581	2 364	2 302	2 323	2 334	2 312
Liberia	122	1970	7	2 583	2 571	2 576	2 500	2 590	2 630
Madagascar	2 428	2004–05	1	35 145	35 390	36 075	36 350	40 843	41 395
Malawi	2 666	2006–07	37	3 200	3 857	3 930	4 320	4 820	5 580
Mali	805	2004–05	1	31 698	31 778	32 083	32 133	39 339	41 621
Mauritania	100	1980	7	39 522	39 493	39 484	39 666	39 712	39 711
Mauritius	..	..		99	112	114	110	102	89
Mozambique	3 065	1999–2000	1	46 649	47 009	47 150	47 730	48 250	49 400
Namibia	102	1996–97	1	38 642	38 653	38 657	38 662	38 820	38 809
Niger	669	1980	7	31 500	31 230	30 280	34 105	38 000	43 782
Nigeria	308	1960	7	68 800	69 900	70 385	72 335	71 900	76 200
Réunion	8	2010	38	61	62	65	63	49	46
Rwanda	1 675	2007–08	39	1 315	1 448	1 760	1 877	1 749	1 920
Sao Tome and Principe	14	1990	7	35	37	37	42	51	49
Senegal	437	1998–99	1	8 647	8 946	8 840	8 709	8 810	9 505
Seychelles	5	2002	1	5	5	5	4	4	3
Sierra Leone	223	1980	7	2 612	2 669	2 729	2 825	2 992	3 435
Somalia	..	..		43 905	43 955	44 005	44 042	44 071	44 129
South Africa	1 093	2000	1	101 335	95 390	94 100	96 005	98 013	96 374
Sudan (former)	..	..		108 840	109 843	110 480	122 965	132 093	
Swaziland	74	1990	7	1 468	1 494	1 284	1 227	1 224	1 222
Togo	430	1996	1	3 070	2 880	3 035	3 195	3 480	3 720
Uganda	3 833	2002	1	9 018	10 030	10 760	12 032	12 612	14 062
United Republic of Tanzania	4 902	2002–03	1	26 000	32 000	33 000	34 003	34 100	37 300
Zambia	1 306	2000	1	19 307	20 053	19 836	20 826	22 555	23 435
Zimbabwe	438	1960	7	10 985	11 835	12 350	13 180	15 240	16 320
HIGH-INCOME COUNTRIES	**21 867**			**1 297 955**	**1 294 798**	**1 282 444**	**1 290 691**	**1 315 429**	**1 246 991**
Andorra	..	..		26	25	21	19	19	20
Aruba	..	..		2	2	2	2	2	2
Australia	141	2001	1	461 585	483 253	482 741	462 974	455 700	409 673
Austria	199	1999–2000	1	4 050	3 894	3 689	3 519	3 376	2 869
Bahamas	2	1994	7	10	10	11	12	13	15

TABLE A1 *(cont.)*

	Number of holdings *(Thousands)*	Census year/ round	Source	Agricultural area *(Thousand ha)*					
				1961	1971	1981	1991	2001	2011
Bahrain	1	1980	7	7	7	9	8	9	8
Barbados	17	1989	7	19	19	19	19	18	15
Belgium	43	2010	40	..	..	..	..	1 389	1 337
Belgium-Luxembourg				1 811	1 756	1 460	1 423	..	..
Bermuda	..	..		1	1	1	1	1	1
Brunei Darussalam	6	1960	7	21	19	14	11	11	11
Canada	247	2001	1	69 825	68 661	65 889	67 753	67 502	62 597
Cayman Islands	..	..		3	3	3	3	3	3
Croatia	450	2003	1					1 178	1 326
Cyprus	39	2010	41	205	235	173	161	140	119
Czech Republic	23	2010	42					4 278	4 229
Czechoslovakia				7 277	7 077	6 843	6 723		
Denmark	58	1999–2000	1	3 160	2 951	2 897	2 770	2 676	2 690
Equatorial Guinea	..	..		314	334	334	334	334	304
Estonia	84	2001	1					890	945
Faroe Islands	..	..		3	3	3	3	3	3
Finland	64	2010	43	2 775	2 700	2 517	2 425	2 222	2 286
France	664	1999–2000	1	34 539	32 623	31 687	30 426	29 631	29 090
French Polynesia	..	..		44	44	44	43	43	46
Germany	472	1999–2000	1	19 375	18 952	18 461	17 136	17 034	16 719
Greece	817	1999–2000	1	8 910	9 155	9 206	9 164	8 502	8 152
Greenland	..	..		235	235	235	236	236	236
Guam	0	2007	44	16	17	20	20	20	18
Hungary	967	2000	1	7 083	6 855	6 601	6 460	5 865	5 337
Iceland	..	..		2 120	1 991	1 900	1 901	1 889	1 591
Ireland	142	2000	1	5 640	5 672	5 732	4 442	4 410	4 555
Israel	..	..		511	527	538	578	561	521
Italy	2 591	2000	1	20 683	17 649	17 551	16 054	15 502	13 933
Japan	3 120	2000	1	7 110	6 541	6 042	5 654	4 793	4 561
Kuwait	..	..		135	135	136	141	151	152
Liechtenstein	..	..		9	9	9	7	7	7
Luxembourg	3	1999–2000	1	..	..	..	..	128	131
Malta	13	2010	45	18	14	13	13	10	10
Monaco	..	..		..	..	..	..	..	..
Netherlands	102	1999–2000	1	2 314	2 128	2 011	1 991	1 931	1 895
New Caledonia	6	2002	1	261	263	265	229	246	251
New Zealand	70	2002	1	15 777	15 670	17 332	16 119	15 418	11 371

TABLE A1 *(cont.)*

	Number of holdings *(Thousands)*	Census year/ round	Source	Agricultural area *(Thousand ha)*					
				1961	1971	1981	1991	2001	2011
Northern Mariana Islands	0	2007	46				4	3	3
Norway	71	1999	1	1 034	931	936	1 010	1 047	998
Oman	..	..		1 035	1 042	1 051	1 080	1 074	1 771
Poland	2 933	2002	1	20 322	19 508	18 910	18 753	17 788	14 779
Portugal	416	1999	1	3 875	3 935	3 982	3 920	3 795	3 636
Puerto Rico	18	2002	1	616	530	467	420	235	190
Qatar	4	2000–01	1	51	51	56	61	66	66
Republic of Korea	3 270	2000	1	2 113	2 299	2 245	2 161	1 945	1 756
Saint Kitts and Nevis	3	2000	1	20	15	15	12	9	6
San Marino	..	..		1	1	1	1	1	1
Saudi Arabia	242	1999	1	86 170	86 467	87 013	123 672	173 791	173 355
Singapore	16	1970	7	14	10	7	1	1	1
Slovakia	71	2001	1					2 255	1 930
Slovenia	75	2010	47					510	459
Spain	1 764	1999	1	33 230	32 684	31 206	30 371	29 520	27 534
Sweden	81	1999–2000	1	4 237	3 758	3 675	3 358	3 154	3 066
Switzerland	108	1990	7	1 736	1 665	1 649	1 601	1 563	1 532
Trinidad and Tobago	19	2004	1	97	101	95	81	60	54
Turks and Caicos Islands	..	..		1	1	1	1	1	1
United Arab Emirates	..	..		208	212	227	310	567	397
United Kingdom	233	1999–2000	1	19 800	18 843	18 320	18 143	16 953	17 164
United States of America	2 205	2007	48	447 509	433 300	428 163	426 948	414 944	411 263
United States Virgin Islands	0	2007	49	12	15	16	10	7	4

TABLE A2
Shares of agricultural holdings and agricultural area, by land size class

		<1 ha	1–2 ha	2–5 ha	5–10 ha	10–20 ha	20–50 ha	>50 ha
		(Percentage)						
LOW-INCOME COUNTRIES	**holdings**	**63**	**20**	**13**	**3**	**1**	**0**	**0**
	area	**20**	**22**	**31**	**16**	**9**	**1**	**2**
LOWER-MIDDLE-INCOME COUNTRIES	**holdings**	**62**	**19**	**14**	**4**	**1**	**0**	**0**
	area	**15**	**16**	**26**	**15**	**9**	**8**	**11**
UPPER-MIDDLE-INCOME COUNTRIES	**holdings**	**27**	**15**	**27**	**13**	**8**	**6**	**5**
	area	**0**	**1**	**3**	**3**	**4**	**7**	**81**
HIGH-INCOME COUNTRIES	**holdings**	**34**	**18**	**15**	**9**	**7**	**7**	**9**
	area	**1**	**1**	**2**	**2**	**4**	**8**	**82**
WORLD	**holdings**	**72**	**12**	**10**	**3**	**1**	**1**	**1**
	area	**8**	**4**	**7**	**5**	**5**	**7**	**65**
LOW- AND MIDDLE-INCOME COUNTRIES								
East Asia and the Pacific								
American Samoa	holdings	57	26	13	3	1	0	..
	area	19	28	30	14	6	3	..
China	holdings	93	5	2	0	0	..	..
	area	..	..	..	..	..	..	..
Cook Islands	holdings	82	14	5	..	..	..	..
	area	43	29	28	..	..	..	..
Fiji	holdings	43	12	20	13	7	3	2
	area	2	3	11	15	14	17	39
Indonesia	holdings	71	17	11	1	0	..	..
	area	30	25	34	8	3	..	..
Lao People's Democratic Republic	holdings	38	35	26	..	..	..	..
	area	13	30	57	..	..	..	..
Myanmar	holdings	34	23	30	11	2	0	..
	area	5	14	37	29	13	3	..
Philippines	holdings	40	28	24	6	2	0	..
	area	9	17	33	20	10	11	..
Samoa	holdings	19	32	30	12	5	2	..
	area	2	11	25	22	18	21	..
Thailand	holdings	20	23	37	16	4	1	0
	area	3	9	34	31	13	5	5
Viet Nam	holdings	85	10	5	0	0	..	..
	area	..	..	..	..	..	..	..
Europe and Central Asia								
Albania	holdings	60	30	10	..	..	..	..
	area	7	11	83	..	..	..	..

TABLE A2 *(cont.)*

		<1 ha	1–2 ha	2–5 ha	5–10 ha	10–20 ha	20–50 ha	>50 ha
		(Percentage)						
Bulgaria	holdings	77	..	20	..	..	2	1
	area	7	..	8	..	..	7	78
Georgia	holdings	70	23	5	1	0	0	0
	area	24	23	12	5	4	4	27
Kyrgyzstan	holdings	85	7	5	2	1	0	0
	area	8	8	15	10	8	9	42
Latvia	holdings	0	6	20	22	24	20	7
	area	..	0	3	8	17	31	40
Lithuania	holdings	0	8	47	23	14	6	2
	area	0	1	14	15	18	17	35
Romania	holdings	50	20	23	6	1	0	0
	area	5	8	20	11	4	2	50
Serbia	holdings	28	19	31	17	5	1	..
	area	5	9	30	33	16	7	..
Turkey	holdings	17	18	31	18	11	5	1
	area	1	4	16	21	24	23	11
Latin America and the Caribbean								
Argentina	holdings	..	..	15	8	10	16	51
	area	..	..	0	0	0	1	98
Brazil	holdings	11	10	16	13	14	17	19
	area	0	0	1	1	3	7	88
Chile	holdings	15	10	18	16	15	14	13
	area	0	0	1	1	3	5	90
Colombia	holdings	18	14	21	14	11	11	11
	area	0	1	3	4	6	14	72
Dominica	holdings	53	21	18	5	1	1	1
	area	8	15	22	14	6	10	25
Ecuador	holdings	29	14	20	12	9	9	6
	area	1	1	4	6	8	19	61
French Guiana	holdings	16	31	42	6	2	2	..
	area	2	9	25	8	4	51	..
Grenada	holdings	85	8	5	1	0	0	..
	area	18	14	20	11	7	30	..
Guadeloupe	holdings	31	27	32	7	2	1	..
	area	5	13	33	16	7	26	..
Guatemala	holdings	78	10	6	2	1	2	0
	area	12	7	10	9	5	36	21
Honduras	holdings	..	..	55	16	12	17	..
	area	..	..	8	7	10	75	..

TABLE A2 *(cont.)*

		<1 ha	1–2 ha	2–5 ha	5–10 ha	10–20 ha	20–50 ha	>50 ha
		(Percentage)						
Jamaica	holdings	69	15	12	2	1	0	0
	area	11	9	16	6	4	6	48
Martinique	holdings	64	13	16	4	2	1	..
	area	9	8	20	11	9	44	..
Nicaragua	holdings	12	9	19	14	15	17	13
	area	0	0	2	4	8	20	66
Panama	holdings	53	10	12	7	6	7	5
	area	1	1	3	4	7	18	67
Paraguay	holdings	10	10	20	22	22	10	7
	area	0	0	1	2	3	4	90
Peru	holdings	..	..	70	15	7	5	3
	area	..	..	5	5	4	8	78
Saint Lucia	holdings	63	18	15	3	1	0	..
	area	31	16	20	4	3	25	..
Saint Vincent and the Grenadines	holdings	73	15	10	2	1	0	..
	area	19	21	25	10	7	18	..
Uruguay	holdings	..	..	11	12	12	16	49
	area	..	..	0	0	1	2	97
Venezuela (Bolivarian Republic of)	holdings	9	14	26	15	12	10	14
	area	0	0	1	2	2	5	89
Middle East and North Africa								
Algeria	holdings	22	13	23	18	14	9	2
	area	1	2	9	14	22	29	23
Egypt	holdings	87	8	4	1	0	0	..
	area	37	18	18	9	6	11	..
Iran (Islamic Republic of)	holdings	47	12	18	11	7	3	1
	area	2	4	13	18	21	21	20
Jordan	holdings	54	32	7	4	2	0	0
	area	4	22	15	15	18	9	17
Lebanon	holdings	73	14	10	2	1	0	0
	area	20	15	25	9	11	11	9
Libya	holdings	14	10	25	23	16	9	1
	area	..	..	..	..	..	..	..
Morocco	holdings	25	18	28	17	8	3	1
	area	2	5	17	22	22	17	15
Yemen	holdings	73	11	9	7	..	..	..
	area	16	10	18	56	..	..	..

TABLE A2 *(cont.)*

		<1 ha	1–2 ha	2–5 ha	5–10 ha	10–20 ha	20–50 ha	>50 ha
		(Percentage)						
South Asia								
India	holdings	63	19	14	3	1	0	..
	area	19	20	31	17	8	5	..
Nepal	holdings	75	17	7	1	0	..	..
	area	39	30	24	5	2	..	..
Pakistan	holdings	36	22	28	9	4	1	0
	area	6	10	28	19	16	12	10
Sub-Saharan Africa								
Burkina Faso	holdings	13	19	41	21	5	..	..
	area	2	7	35	37	19	..	..
Côte d'Ivoire	holdings	42	14	19	13	8	3	..
	area	5	5	15	22	27	25	..
Democratic Republic of the Congo	holdings	87	10	3	..	..	..	..
	area	63	23	14	..	..	..	..
Ethiopia	holdings	63	24	12	1	0	..	..
	area	27	33	33	6	1	..	..
Guinea	holdings	34	31	28	7	..	..	..
	area	10	22	42	26	..	..	..
Guinea-Bissau	holdings	70	18	10	2	0	..	..
	area	..	..	..	..	..	..	..
Lesotho	holdings	47	29	20	4	..	..	..
	area	..	..	..	..	..	..	..
Malawi	holdings	78	17	5	..	..	..	..
	area	..	..	..	..	..	..	..
Mozambique	holdings	54	30	14	2	0	0	0
	area	..	..	..	..	..	..	..
Namibia	holdings	14	25	49	11	1	0	0
	area	3	13	54	25	4	1	0
Réunion	holdings	24	18	29	21	5	2	..
	area	2	5	20	30	15	29	..
Senegal	holdings	21	17	33	21	8	1	..
	area	2	6	25	34	24	9	..
Uganda	holdings	49	24	17	6	4	..	..
	area	11	16	25	18	30	..	..
HIGH-INCOME COUNTRIES								
Austria	holdings	..	15	22	19	22	18	4
	area	..	2	5	10	18	24	41

TABLE A2 *(cont.)*

		<1 ha	1–2 ha	2–5 ha	5–10 ha	10–20 ha	20–50 ha	>50 ha
		(Percentage)						
Bahamas	holdings	36	25	20	8	4	3	3
	area	1	3	5	4	5	7	74
Barbados	holdings	95	3	1	0	0	0	1
	area	10	3	3	1	2	3	78
Belgium	holdings	..	17	14	13	16	27	12
	area	..	1	2	4	11	39	43
Canada	holdings	..	2	3	4	5	14	72
	area	..	..	..	..	..	..	..
Croatia	holdings	51	16	19	9	4	1	..
	area	6	7	20	21	15	31	..
Cyprus	holdings	55	17	16	6	3	2	1
	area	6	7	14	13	14	16	30
Czech Republic	holdings	29	15	17	11	9	8	10
	area	0	0	1	1	2	4	92
Denmark	holdings	..	2	2	16	20	30	31
	area	..	0	0	3	6	21	70
Estonia	holdings	20	20	24	16	11	6	3
	area	1	2	6	9	12	14	56
Finland	holdings	..	3	7	14	25	37	14
	area	..	1	3	7	19	43	28
France	holdings	..	17	12	9	11	21	30
	area	..	1	1	2	4	17	75
French Polynesia	holdings	77	12	6	2	1	2	..
	area	8	5	6	5	5	71	..
Germany	holdings	..	8	17	16	19	24	17
	area	..	0	2	4	8	22	63
Greece	holdings	..	49	28	13	6	3	1
	area	..	11	21	20	19	18	10
Guam	holdings	30	16	27	16	7	5	..
	area	3	4	18	21	18	36	..
Hungary	holdings	27	13	19	11	14	10	6
	area	..	..	..	..	..	..	..
Ireland	holdings	..	2	6	12	24	39	17
	area	..	0	1	3	12	40	45
Italy	holdings	38	19	21	10	6	4	2
	area	2	4	9	9	11	16	49
Japan	holdings	68	20	9	1	1	0	0
	area	25	23	22	8	7	10	5
Luxembourg	holdings	..	12	10	10	7	19	42
	area	..	0	1	2	3	15	79

TABLE A2 *(cont.)*

		<1 ha	1–2 ha	2–5 ha	5–10 ha	10–20 ha	20–50 ha	>50 ha
		(Percentage)						
Malta	holdings	76	15	8	1	0	..	..
	area	33	25	29	10	3	..	..
Netherlands	holdings	..	16	15	16	17	28	8
	area	..	1	3	6	12	43	36
New Zealand	holdings	..	..	17	10	10	14	48
	area	..	..	..	..	..	..	..
Northern Mariana Islands	holdings	26	28	28	8	4	7	..
	area	3	7	17	12	12	48	..
Norway	holdings	2	4	15	24	32	22	2
	area	0	0	4	12	31	43	10
Poland	holdings	33	18	21	15	9	3	1
	area	3	5	13	18	21	16	25
Portugal	holdings	27	28	24	10	6	3	2
	area	3	6	10	9	10	10	52
Puerto Rico	holdings	..	..	53	20	13	9	6
	area	..	..	7	9	11	17	56
Qatar	holdings	69	5	6	4	4	6	5
	area	1	1	2	2	5	16	73
Republic of Korea	holdings	59	31	10	..	..	..	..
	area	31	41	28	..	..	..	..
Slovakia	holdings	70	12	10	2	1	1	3
	area	..	..	..	..	..	..	..
Slovenia	holdings	28	13	23	18	13	5	..
	area	..	..	..	..	..	..	..
Spain	holdings	26	15	22	13	10	8	7
	area	..	..	..	..	..	..	..
Saint Kitts and Nevis	holdings	..	96	3	0	1	..	..
	area	..	..	..	..	..	..	..
Sweden	holdings	..	3	9	17	21	27	23
	area	..	2	4	9	14	25	47
Switzerland	holdings	20	7	11	14	29	18	1
	area	1	1	3	9	36	43	7
Trinidad and Tobago	holdings	35	18	34	9	3	1	0
	area	3	5	22	14	6	8	42
United Kingdom	holdings	..	14	9	11	13	21	32
	area	..	0	1	1	3	10	85
United States of America	holdings	..	..	11	10	14	22	44
	area	..	..	0	0	1	4	94
United States Virgin Islands	holdings	..	50	23	13	4	7	4
	area	..	2	3	5	2	12	75

TABLE A3
Average level and rate of change in agricultural labour productivity, 1961–2012

	Agricultural labour productivity (value of agricultural production/agricultural worker)									
	Average annual level (Constant 2004–06 international dollars)					Average annual rate of change (Percentage)				
	1961–1971	1971–1981	1981–1991	1991–2001	2001–2012	1961–1971	1971–1981	1981–1991	1991–2001	2001–2012
LOW-INCOME COUNTRIES	**405**	**412**	**416**	**419**	**490**	**0.8**	**0.3**	**–0.2**	**0.7**	**1.9**
LOWER-MIDDLE-INCOME COUNTRIES	**748**	**848**	**937**	**902**	**1 057**	**2.0**	**0.7**	**1.4**	**0.5**	**2.3**
UPPER-MIDDLE-INCOME COUNTRIES	**527**	**609**	**720**	**1 003**	**1 454**	**2.2**	**1.6**	**1.3**	**3.7**	**3.5**
HIGH-INCOME COUNTRIES	**5 556**	**8 627**	**12 211**	**18 095**	**27 112**	**4.7**	**4.2**	**3.2**	**4.5**	**3.7**
WORLD	**943**	**1 059**	**1 141**	**1 261**	**1 535**	**1.7**	**1.0**	**0.4**	**1.7**	**2.1**
LOW- AND MIDDLE-INCOME COUNTRIES	**596**	**671**	**755**	**879**	**1 144**	**1.9**	**1.0**	**1.2**	**2.2**	**2.8**
East Asia and the Pacific	**306**	**353**	**446**	**621**	**921**	**2.3**	**1.6**	**2.0**	**4.1**	**3.6**
American Samoa	695	474	304	282	529	–1.2	–2.7	–4.9	4.9	4.6
Cambodia	488	266	350	423	601	1.1	–4.7	3.4	2.2	6.3
China, mainland	253	290	379	567	869	2.9	1.2	2.6	5.0	3.8
Democratic People's Republic of Korea	512	736	918	946	1 131	2.1	4.3	1.9	–1.3	0.9
Fiji	2 068	1 887	1 984	1 867	1 696	0.7	1.7	–0.1	–1.4	–1.3
Indonesia	426	530	665	783	1 035	2.1	2.2	1.5	0.6	3.8
Kiribati	1 647	1 554	1 694	1 620	2 189	–0.8	1.8	–2.3	2.4	3.6
Lao People's Democratic Republic	331	325	388	443	623	3.1	0.7	0.0	3.6	2.0
Malaysia	1 315	2 056	3 202	4 748	7 827	4.4	3.7	5.1	3.1	5.2
Marshall Islands	..	..	363	391	563	..	..	..	–14.5	13.7
Micronesia (Federated States of)	..	..	..	752	894	..	..	..	..	1.9
Mongolia	2 959	3 326	3 441	3 318	3 195	0.6	0.8	0.7	0.9	3.5
Myanmar	342	355	417	443	723	–0.4	2.5	–2.6	3.5	4.7
Palau	..	..	..	..	..	..	..	..	..	..
Papua New Guinea	1 046	1 211	1 220	1 216	1 258	1.7	1.1	–0.8	0.4	0.4
Philippines	800	970	1 036	1 125	1 380	0.8	3.1	0.0	0.6	2.4
Samoa	1 646	1 797	1 989	1 774	2 551	–0.6	1.9	–1.4	3.5	3.4
Solomon Islands	725	780	829	726	772	–0.3	2.6	–3.4	–0.7	2.3
Thailand	591	725	826	1 052	1 448	1.4	3.3	0.5	2.6	3.2
Timor-Leste	502	466	425	415	402	0.7	–1.9	–0.4	0.1	–1.1
Tonga	2 164	2 316	2 134	1 914	2 143	–1.6	2.9	–3.0	0.1	1.6
Tuvalu	651	609	644	753	857	–1.6	6.6	–0.9	0.2	1.5
Vanuatu	2 004	2 015	2 131	1 980	1 799	–0.1	2.7	–1.1	0.2	1.7
Viet Nam	317	335	420	547	820	–0.3	1.2	1.3	4.1	3.2

TABLE A3 *(cont.)*

	Agricultural labour productivity (value of agricultural production/agricultural worker)									
	Average annual level (*Constant 2004–06 international dollars*)					Average annual rate of change (*Percentage*)				
	1961–1971	1971–1981	1981–1991	1991–2001	2001–2012	1961–1971	1971–1981	1981–1991	1991–2001	2001–2012
Europe and Central Asia	**1 928**	**2 775**	**3 366**	**3 430**	**4 697**	**5.1**	**2.2**	**2.0**	**0.1**	**4.1**
Albania	574	715	736	1 060	1 592	1.9	2.2	–1.4	4.9	4.5
Armenia				2 752	5 271				3.6	7.0
Azerbaijan				1 431	1 939				–0.8	3.5
Belarus				4 933	9 253				1.4	8.4
Bosnia and Herzegovina				4 757	14 173				6.0	12.6
Bulgaria	2 216	4 064	6 852	10 057	17 858	7.9	5.9	4.0	6.2	7.0
Georgia				1 847	2 047				3.1	–1.5
Kazakhstan				3 900	5 342				–2.4	3.8
Kyrgyzstan				2 347	2 965				3.4	1.1
Latvia				4 393	5 941				–4.0	6.6
Lithuania				5 513	10 896				1.1	8.8
Montenegro					4 187					..
Republic of Moldova				3 199	5 420				..	5.1
Romania	1 085	2 023	3 005	3 720	7 558	5.2	6.4	1.5	4.0	6.5
Russian Federation				4 194	5 731				..	4.1
Serbia					5 970					..
Serbia and Montenegro				3 768					2.6	
Tajikistan				1 275	1 387				–2.0	0.0
The former Yugoslav Republic of Macedonia				4 930	8 677				5.3	7.7
Turkey	1 562	2 053	2 328	2 739	3 789	2.5	3.0	0.4	2.5	4.2
Turkmenistan				2 375	3 153				–0.6	1.2
Ukraine				4 104	6 472				–0.1	5.8
USSR	2 375	3 293	3 809			5.7	0.7	2.5		
Uzbekistan				2 601	3 228				–0.8	3.7
Yugoslav SFR	891	1 583	2 879			4.6	7.4	4.9		
Latin America and the Caribbean	**2 061**	**2 486**	**3 123**	**4 032**	**5 923**	**1.9**	**2.5**	**2.2**	**3.2**	**3.8**
Antigua and Barbuda	1 057	761	1 112	1 287	1 221	–7.6	2.5	2.7	–0.6	–0.7
Argentina	10 709	14 047	15 802	18 960	25 970	2.8	4.0	–0.1	3.2	3.0
Belize	2 591	3 685	4 266	5 609	5 697	5.4	2.9	0.9	2.1	–2.1
Bolivia (Plurinational State of)	879	1 144	1 194	1 362	1 530	2.6	1.0	1.5	0.8	1.2
Brazil	1 648	2 155	3 383	5 252	9 832	2.0	3.4	5.0	4.6	6.2
Chile	3 111	3 546	4 031	5 631	7 526	2.6	2.0	1.4	3.4	2.4
Colombia	1 622	1 979	2 296	2 872	3 524	1.7	2.2	3.1	1.2	2.0

TABLE A3 *(cont.)*

	Agricultural labour productivity (value of agricultural production/agricultural worker)									
	Average annual level *(Constant 2004–06 international dollars)*					Average annual rate of change *(Percentage)*				
	1961–1971	1971–1981	1981–1991	1991–2001	2001–2012	1961–1971	1971–1981	1981–1991	1991–2001	2001–2012
Costa Rica	2 556	3 796	4 222	6 327	7 991	5.8	1.0	3.9	2.5	2.9
Cuba	3 357	4 128	5 021	3 921	4 503	3.6	3.4	0.5	0.9	0.2
Dominica	2 627	2 771	4 064	4 552	4 051	4.4	0.4	6.6	–1.3	2.1
Dominican Republic	1 990	2 547	2 788	3 039	4 907	0.5	2.2	–0.6	2.9	5.6
Ecuador	2 194	2 279	2 557	3 616	4 693	0.7	1.1	2.2	3.1	2.7
El Salvador	1 130	1 296	1 223	1 340	1 606	–0.6	2.0	0.0	0.9	3.1
Grenada	1 678	1 890	1 874	1 849	1 536	5.6	2.2	–0.3	–1.8	–2.2
Guatemala	910	1 177	1 207	1 635	1 873	2.1	2.4	0.5	4.0	1.9
Guyana	3 518	3 716	3 338	5 133	6 078	1.0	0.0	–1.9	4.9	1.1
Haiti	455	535	551	452	440	1.5	1.2	–1.4	–0.6	0.1
Honduras	1 211	1 419	1 526	1 710	2 548	4.5	0.8	0.8	0.6	4.3
Jamaica	1 578	1 548	1 481	2 123	2 443	2.2	–2.4	2.9	1.8	1.2
Mexico	1 656	2 021	2 390	2 803	3 797	3.0	2.0	0.5	2.9	2.6
Nicaragua	1 794	2 305	1 747	1 974	3 540	4.3	–0.1	–2.5	4.7	5.5
Panama	2 291	3 119	3 162	2 901	3 286	4.7	2.4	–1.7	0.8	2.0
Paraguay	2 239	2 558	3 303	3 763	4 744	0.7	2.4	3.5	0.3	3.9
Peru	1 338	1 349	1 304	1 401	2 000	1.4	–1.3	–0.6	4.1	3.7
Saint Lucia	3 396	3 112	3 603	3 211	1 337	1.8	–1.5	4.5	–9.9	–5.1
Saint Vincent and the Grenadines	1 821	1 885	2 492	2 321	2 023	0.0	0.6	3.7	–4.3	0.3
Suriname	2 242	3 453	4 375	3 539	2 923	5.5	5.9	–2.4	–3.6	1.2
Uruguay	8 216	9 214	10 828	12 825	17 440	1.9	1.7	0.2	2.6	5.5
Venezuela (Bolivarian Republic of)	2 491	3 640	4 560	5 722	7 756	4.6	4.0	1.1	3.6	2.7
Middle East and North Africa	**1 032**	**1 284**	**1 703**	**2 359**	**2 993**	**2.2**	**2.0**	**3.5**	**2.2**	**2.1**
Algeria	978	1 071	1 323	1 424	1 726	1.4	0.5	2.8	–1.4	4.0
Djibouti	195	178	242	192	244	–1.4	0.7	1.7	0.1	2.6
Egypt	887	983	1 233	2 179	3 051	1.7	0.7	5.0	4.3	2.8
Iran (Islamic Republic of)	1 054	1 514	2 102	3 047	3 622	3.4	3.2	2.4	2.1	1.3
Iraq	1 349	1 874	3 179	4 172	5 385	2.5	4.5	4.6	4.2	2.0
Jordan	3 066	2 556	4 590	5 684	8 886	–8.7	7.5	3.5	1.3	4.1
Lebanon	2 808	4 647	10 519	25 410	35 787	7.3	2.6	11.7	3.9	3.9
Libya	1 144	2 436	4 585	8 286	13 778	8.0	6.5	6.7	4.8	6.3
Morocco	858	917	1 222	1 508	2 319	3.6	–1.0	6.5	1.1	5.1
Occupied Palestinian Territory	..	..	..	3 687	4 977	..	..	..	..	0.2
Syrian Arab Republic	2 122	3 134	4 069	4 104	4 820	–0.8	8.2	–3.3	3.1	–1.1

TABLE A3 *(cont.)*

	Agricultural labour productivity (value of agricultural production/agricultural worker)									
	Average annual level (*Constant 2004–06 international dollars*)					Average annual rate of change (*Percentage*)				
	1961–1971	1971–1981	1981–1991	1991–2001	2001–2012	1961–1971	1971–1981	1981–1991	1991–2001	2001–2012
Tunisia	1 562	2 361	2 891	3 671	4 163	3.4	0.2	5.3	–0.4	2.3
Yemen	422	500	547	545	717	–1.3	2.4	1.1	1.4	3.4
South Asia	**446**	**484**	**562**	**668**	**775**	**0.8**	**1.1**	**1.8**	**1.6**	**2.5**
Afghanistan	736	775	791	694	603	1.4	1.0	–0.5	–1.3	–0.1
Bangladesh	330	324	333	378	537	0.3	1.2	0.2	2.9	3.6
Bhutan	628	593	621	717	526	0.1	–0.6	–0.2	–0.5	–1.4
India	434	474	555	658	763	0.7	1.1	1.8	1.5	2.7
Maldives	317	399	519	511	442	2.3	2.6	0.1	–0.2	–1.1
Nepal	319	332	393	445	457	0.3	0.4	3.0	0.1	0.5
Pakistan	826	916	1 133	1 460	1 477	2.4	0.3	4.2	1.0	0.4
Sri Lanka	555	586	619	608	654	0.5	2.2	–1.9	0.5	1.9
Sub-Saharan Africa	**566**	**583**	**581**	**626**	**696**	**1.2**	**–0.2**	**0.8**	**0.8**	**0.8**
Angola	495	413	269	279	467	1.9	–6.9	–1.4	2.4	4.9
Benin	462	543	658	831	1 046	1.9	1.7	2.0	3.9	1.4
Botswana	856	951	975	903	830	3.0	–1.3	0.9	–4.6	2.4
Burkina Faso	210	208	270	334	370	2.0	1.3	3.9	0.4	–0.7
Burundi	452	453	413	350	282	0.8	–0.4	–0.5	–2.5	–2.8
Cameroon	518	649	687	755	1 074	2.7	1.0	0.1	1.7	5.6
Cape Verde	362	306	541	825	1 243	–2.3	5.5	8.4	3.7	5.5
Central African Republic	398	481	502	584	708	2.0	1.3	0.5	2.7	1.7
Chad	585	502	458	463	477	–0.9	–0.1	–0.4	1.1	–0.3
Comoros	439	416	377	391	348	0.5	–1.3	1.0	–0.8	–1.1
Congo	473	444	465	499	679	0.5	–0.3	0.1	2.2	3.8
Côte d'Ivoire	981	1 214	1 334	1 588	1 959	2.3	2.4	0.9	3.1	2.1
Democratic Republic of the Congo	458	449	467	401	297	–0.2	–0.6	0.8	–4.4	–1.2
Eritrea				171	145				0.8	–0.5
Ethiopia				216	265				0.9	2.6
Ethiopia PDR	328	296	272			–0.1	0.1	–2.4		
Gabon	490	633	835	1 011	1 244	2.1	3.5	2.7	1.5	3.0
Gambia	569	441	316	220	223	0.4	–6.5	–5.3	2.3	–1
Ghana	808	723	615	841	1 010	1.0	–5.0	2.6	1.6	1.8
Guinea	401	409	398	400	444	0.3	0.2	–0.2	0.0	1.0
Guinea-Bissau	366	343	408	468	581	–2.9	1.0	1.5	2.3	2.5

TABLE A3 *(cont.)*

	Agricultural labour productivity (value of agricultural production/agricultural worker)									
	Average annual level (*Constant 2004–06 international dollars*)					Average annual rate of change (*Percentage*)				
	1961–1971	1971–1981	1981–1991	1991–2001	2001–2012	1961–1971	1971–1981	1981–1991	1991–2001	2001–2012
Kenya	448	483	500	452	513	0.5	0.5	0.8	–1.5	2.6
Lesotho	429	445	418	384	378	1.6	0.1	–1.7	1.5	–0.1
Liberia	527	597	565	456	480	2.4	–0.5	–2.3	4.1	–1.7
Madagascar	652	649	596	519	446	0.6	–1.0	–0.8	–2.0	0.2
Malawi	267	327	319	344	494	2.0	0.8	–1.6	5.9	3.9
Mali	563	595	727	851	1 088	1.9	2.6	2.1	1.5	3.1
Mauritania	682	603	680	675	632	0.3	1.4	1.6	–0.9	–0.7
Mauritius	2 231	2 291	2 678	3 621	5 016	0.3	–1.2	3.7	2.5	3.0
Mozambique	285	268	202	210	267	1.3	–4.1	–0.7	4.2	3.1
Namibia	2 056	2 343	1 801	1 638	1 655	2.6	–1.7	–1.3	–1.9	0.1
Niger	595	499	446	488	617	–0.2	1.3	–1.3	1.7	1.4
Nigeria	729	721	977	1 793	2 502	1.5	0.3	6.4	4.0	2.0
Rwanda	374	419	418	375	418	2.9	0.9	–1.4	–2.5	3.5
Sao Tome and Principe	1 051	883	598	758	886	1.6	–5.4	–2.7	5.3	–0.6
Senegal	530	416	370	337	328	–3.0	–2.2	0.0	0.4	1.7
Seychelles	375	285	255	258	172	–0.7	–2.9	–1.7	1.3	–3.5
Sierra Leone	351	389	389	374	617	2.4	0.3	0.0	–1.7	8.0
Somalia	865	853	794	713	689	1.8	–2.8	0.6	1.0	–0.2
South Africa	2 602	3 849	4 883	5 688	8 691	2.4	5.6	1.7	2.9	4.7
Sudan	699	828	822	1 027	1 285	1.7	1.2	–0.3	3.2	–0.3
Swaziland	988	1 517	1 941	1 716	1 953	4.4	4.0	0.2	–1.3	2.1
Togo	501	461	458	548	586	0.4	–0.2	1.1	1.9	1.3
Uganda	611	659	502	504	517	3.0	–4.9	–0.2	0.5	–1.1
United Republic of Tanzania	359	372	375	334	411	0.6	1.0	–0.4	–0.6	2.1
Zambia	325	390	337	320	404	1.5	–0.8	0.7	0.1	4.2
Zimbabwe	561	670	570	513	481	1.6	–1.3	–0.7	3.0	–1.2
HIGH-INCOME COUNTRIES	**5 556**	**8 627**	**12 211**	**18 095**	**27 112**	**4.7**	**4.2**	**3.2**	**4.5**	**3.7**
Andorra	..	..	..	..	..	..	..	..	..	..
Aruba	..	..	..	..	..	..	..	..	..	..
Australia	25 721	33 684	36 881	48 040	51 981	3.4	1.7	0.9	4.1	0.0
Austria	5 390	9 084	12 743	17 365	25 584	6.4	4.7	1.9	4.7	3.8
Bahamas	1 616	3 490	3 184	3 956	5 765	8.6	1.8	–1.0	6.8	3.6
Bahrain	1 938	3 948	4 437	6 611	6 756	3.6	8.8	4.9	4.5	1.6
Barbados	3 545	4 481	5 362	6 644	9 319	3.4	4.8	1.2	3.3	3.6

TABLE A3 *(cont.)*

	Agricultural labour productivity (value of agricultural production/agricultural worker)									
	Average annual level (Constant 2004–06 international dollars)					Average annual rate of change (Percentage)				
	1961–1971	1971–1981	1981–1991	1991–2001	2001–2012	1961–1971	1971–1981	1981–1991	1991–2001	2001–2012
Belgium	..	..	..	..	81 004	..	..	..	..	0.8
Belgium-Luxembourg	17 118	31 159	43 511	63 982		7.1	4.3	3.1	3.6	
Bermuda	2 613	1 728	1 870	1 942	1 984	–0.6	–1.7	1.9	–1.0	1.4
Brunei Darussalam	1 027	2 029	3 984	13 327	30 608	4.5	6.2	2.3	19.5	2.7
Canada	13 527	16 925	26 208	47 408	68 306	4.7	1.1	6.3	4.8	3.7
Cayman Islands	191	197	153	65	44	..	0.5	–13.5	0.1	–6.4
China, Hong Kong SAR	..	3 998	4 776	3 790	5 523	..	..	–1.3	5.1	–1.2
China, Macao SAR	..	329	681	..	..	..	..	..	..	..
Croatia				5 348	11 331				7.9	7.4
Cyprus	2 752	3 512	5 958	9 559	11 229	7.2	1.6	6.1	4.0	0.1
Czech Republic				8 394	10 133				1.3	1.9
Czechoslovakia	3 349	5 292	7 139			5.5	3.3	2.4		
Denmark	13 504	20 015	29 926	44 715	69 608	2.9	5.4	2.7	4.6	4.2
Equatorial Guinea	553	366	338	293	268	–1.1	0.9	–1.6	–2.0	0.2
Estonia				4 888	6 686				–2.8	5.7
Faroe Islands	675	1 701	1 771	1 875	1 859	29.7	–0.3	0.8	0.0	0.2
Finland	3 720	5 386	8 008	11 312	17 191	3.3	4.3	3.2	3.8	3.4
France	8 651	14 776	23 992	38 045	57 626	5.6	5.1	4.1	4.7	4.2
French Polynesia	1 192	857	665	605	721	–3.1	–1.8	–2.0	0.1	2.1
Germany	6 538	10 827	17 267	24 652	41 180	7.5	3.3	5.0	5.0	5.0
Greece	2 740	4 642	6 963	9 557	11 048	4.7	5.1	3.4	1.9	0.8
Greenland	957	905	1 342	1 257	1 260	5.8	4.2	–0.2	–2.6	..
Guam	313	404	398	425	512	2.4	3.6	–2.0	3.3	0.4
Hungary	2 975	5 562	9 036	10 544	14 689	5.8	6.0	3.3	3.8	1.8
Iceland	5 701	7 380	6 845	6 069	8 419	0.3	3.7	–4.1	2.3	3.0
Ireland	7 035	12 426	19 236	26 007	27 945	5.8	5.4	4.5	1.5	1.0
Israel	9 749	17 752	25 417	31 466	48 546	6.8	4.4	2.0	3.4	3.5
Italy	5 208	8 795	12 807	20 424	31 185	6.9	5.0	2.8	5.0	3.6
Japan	1 265	2 381	3 837	5 619	10 159	6.7	6.5	3.6	4.5	6.5
Kuwait	7 120	6 232	8 620	10 185	15 137	–2.4	–0.1	–1.1	18.2	1.5
Liechtenstein	1 869	2 227	3 856	..	..	0.8	5.4	3.5	..	..
Luxembourg	..	..	..	..	54 859	..	..	..	..	2.4
Malta	4 359	5 643	10 808	25 729	37 968	5.6	–1.3	13.1	3.3	0.5
Monaco	..	..	..	..	..	..	..	..	..	..
Netherlands	17 006	29 357	37 734	42 513	53 204	6.9	3.8	0.5	0.9	4.1

TABLE A3 *(cont.)*

	Agricultural labour productivity (value of agricultural production/agricultural worker)									
	Average annual level *(Constant 2004–06 international dollars)*					Average annual rate of change *(Percentage)*				
	1961–1971	1971–1981	1981–1991	1991–2001	2001–2012	1961–1971	1971–1981	1981–1991	1991–2001	2001–2012
New Caledonia	1 125	815	681	664	698	–1.9	–3.9	–2.6	0.6	0.2
New Zealand	37 078	40 502	41 093	45 780	53 997	2.7	0.2	–0.4	1.9	1.0
Northern Mariana Islands	..	..	..	..	..	..	..	..	..	..
Norway	4 729	6 849	8 726	10 717	13 379	4.6	3.0	1.8	2.0	2.3
Oman	410	550	765	828	1 073	1.6	4.2	–1.3	5.1	0.9
Poland	2 076	2 791	3 307	3 727	5 192	2.0	3.1	2.2	2.4	3.5
Portugal	2 498	2 887	3 582	5 338	7 140	3.3	–1.3	6.0	2.8	3.2
Puerto Rico	5 077	6 677	8 398	10 075	17 075	1.1	5.3	2.0	2.2	6.8
Qatar	1 763	2 210	3 673	8 148	7 979	1.4	13.4	–0.3	7.3	–5.6
Republic of Korea	621	954	1 726	3 572	6 640	3.5	5.4	7.4	7.3	5.8
San Marino	..	..	..	..	..	..	..	..	..	..
Saudi Arabia	457	646	1 578	3 283	5 712	2.0	3.3	12.7	5.5	5.0
Singapore	4 924	13 566	18 956	12 479	11 452	10.7	6.1	2.3	–8.0	5.6
Slovakia				6 663	7 181				–0.6	1.0
Slovenia				26 890	72 075				11.7	8.5
Spain	3 170	6 050	10 416	17 341	26 703	4.9	6.5	5.1	6.2	2.5
Sweden	6 833	9 687	12 864	17 030	22 194	3.2	3.6	1.2	4.4	1.6
Switzerland	8 593	11 895	13 495	13 631	16 786	3.7	3.2	–0.8	1.7	2.4
Trinidad and Tobago	2 773	3 092	2 641	2 738	3 092	2.0	0.3	–0.3	0.0	–0.5
Turks and Caicos Islands	..	..	..	..	..	..	..	..	..	..
United Arab Emirates	3 708	3 607	3 207	6 838	5 382	4.2	–2.8	–3.4	12.2	–10.1
United Kingdom	14 465	20 049	25 218	30 203	32 257	4.2	2.1	1.7	0.8	1.4
United States Virgin Islands	546	232	218	193	268	–14.9	0.1	–4.5	1.2	4.2
United States of America	23 145	33 130	38 423	52 615	74 723	4.6	2.5	1.1	3.6	3.4

- **References**
- **Special chapters of**
  The State of Food and Agriculture

References

Adekunle, A. & Fatunabi, A. 2012. Approaches for setting-up multi-stakeholder platforms for agricultural research and development. *World Applied Sciences Journal,* 16(7), pp. 981–988.

Adekunle, A., Ellis-Jones, J., Ajibefun, I., Nyikal, R.A., Bangali, S., Fatunbi, O. & Ange, A. 2012. *Agricultural innovation in sub-Saharan Africa: experiences from multiple-stakeholder approaches.* Accra Forum for Agricultural Research in Africa (FARA).

Adeleke, O.A., Adesiyan, O.I., Olaniyi, O.A., Adelalu, K.O. & Matanmi, H.M. 2008. Gender differentials in the productivity of cereal crop farmers: a case study of maize farmers in Oluyole local government area of Oyo State. *Agricultural Journal,* 3(3): 193–198.

Adhiguru, P., Birthal, P. & Ganesh Kumar, B. 2009. Strengthening pluralistic agricultural information delivery systems in India. *Agricultural Economics Research Review,* 22(Jan.–June), pp. 71–79.

Akresh, R. 2008. *(In)Efficiency in intrahousehold allocations.* Working Paper. Department of Economics. Urbana, USA, University of Illinois at Urbana Champaign.

Alexandratos, N. & Bruinsma, J. 2012. *World agriculture towards 2030/2050: the 2012 revision.* Rome, FAO.

Ali, D. & Deininger, K. 2014, February. *Is there a farm-size productivity relationship in African agriculture? Evidence from Rwanda.* World Bank Policy Research Working Paper No. 6770. Washington, DC, World Bank.

Alston, J., Beddow, J. & Pardey, P. 2010. Global patterns of crop yields and other partial productivity measures and prices. *In* J. Alston, B. Babcock & P. Pardey, eds. *The shifting patterns of agricultural production and productivity worldwide.* Ames, Iowa, USA, The Midwest Agribusiness Trade Research and Information Center.

Alston, J., Marra, M., Pardey, P. & Wyatt, T. 2000. Research returns redux: a meta-analysis of the returns to agricultural R&D. *The Australian Journal of Agricultural and Resource Economics,* 44(2): 185–215.

Amanor, K. & Farrington, J. 1991. NGOs and agricultural technology development. *In* W. Rivera & D. Gustafson, eds. *Agricultural extension: worldwide institutional evolution and forces for change.* Amsterdam, Elsevier.

Anandajayasekeram, P. 2011. *The role of agricultural R&D within the agricultural innovation systems framework.* Conference Working Paper 6. Prepared for the Agricultural Science and Technology Indicators (ASTI), IFPRI, Forum for Agricultural Research in Africa (FARA) Conference, Accra, 5–7 December 2011.

Anderson, J. 2008. *Agricultural advisory services.* Background paper for the World Development Report 2008, Washington, DC, World Bank.

Anderson, J. & Feder, G. 2007. Agricultural extension. *In* R.A. Evenson & P. Pingali, eds. *Handbook of agricultural economics.* Volume 3. *Agricultural development: farmers, farm production and farm markets,* Chapter 44, pp. 2343–2378, Amsterdam, North Holland.

Arias, P., Hallam, D., Krivonos, E. & Morrison, J. 2013. *Smallholder integration in changing food markets.* Rome, FAO.

Arslan, A., McCarthy, N., Lipper, L., Asfaw, S. & Cattaneo, A. 2013. *Adoption and intensity of adoption of conservation farming practices in Zambia*. ESA Working Paper No. 13-01. Rome, FAO.

Asenso-Okyere, K. & Mekonnen, D. 2012. *The importance of ICTs in the provision of information for improving agricultural productivity and rural incomes in Africa.* UNDP Working Paper 2012-015. New York, USA, UNDP Regional Bureau for Africa.

Asfaw, S., McCarthy, N., Lipper, L., Arslan, A. & Cattaneo, A. 2014. *Climate variability, adaptation strategies and food security in Malawi*. ESA Working Paper No. 14-08, Rome, FAO.

Ashby, J. 2009. The impact of participatory plant breeding. *In* E.G.S. Ceccarelli, ed. *Plant breeding and farmer participation.* Rome, FAO.

Barrett, C. 2008. Smallholder market participation: concepts and evidence from eastern and southern Africa. *Food Policy,* 33(4): 299–317.

Barrett, C., Bellemare, M. & Hou, J. 2010. Reconsidering conventional explanations of the inverse productivity-size relationship. *World Development,* 38(1): 88–97.

Beintema, N. & Di Marcantonio, F. 2009. *Women's participation in agricultural research and higher* *education: key trends in sub-Saharan Africa.* Washington, DC and Nairobi, IFPRI and CGIAR Gender & Diversity Program.

Beintema, N. & Stads, G. 2011. *African agricultural R&D in the new millennium: progress for some, challenges for many.* Washington, DC and Rome, IFPRI and ASTI.

Beintema, N., Stads, G., Fuglie, K. & Heisey, P. 2012. *ASTI global assessment of agricultural R&D spending: developing countries accelerate investment.* Washington, DC and Rome, IFPRI, ASTI and GFAR.

Benin, S., Nkonya, E., Okecho, G., Randriamamonjy, J., Kato, E., Lubadde, G., Kyotalimye, M. & Byekwaso, F. 2011. *Impact of Uganda's national agricultural advisory services program.* Washington, DC, IFPRI.

Benson, A. & Jafry, T. 2013. The state of agricultural extension: an overview and new caveats for the future. *The Journal of Agricultural Education and Extension,* 19(4): 381–393.

Bienabe, C. & Le Coq, L. 2004. *Linking smallholder farmers to markets. Lessons learned from literature review and analytical review of selected projects.* Washington, DC, World Bank.

Birner, R. & Anderson, J. 2007. *How to make agricultural extension demand-driven? The case of India's agricultural extension policy.* Washington, DC, IFPRI.

Birner, R., Davis, K., Pender, J., Nkonya, E., Anandajayasekeram, P., Ekboir, J., Mbabu, A., Spielman, D., Horna, D., Benin, S. & Cohen, M. 2009. From best practice to best fit: a framework for designing and analyzing pluralistic agricultural advisory services. *Journal of agricultural education and extension,* 15(4):341–355.

Branca, G., McCarthy, N., Lipper, L. & Jolejole, M. 2011. *Climate-smart agriculture: a synthesis of empirical evidence of food security and mitigation benefits from improved cropland management.* FAO Mitigation of Climate Change in Agriculture Series No. 3. Rome, FAO.

Braun, A., Jiggins, J., Rölling, N., van den Berg, H. & Snijders, P. 2006. *A global survey and review of farmer field school experiences.* Wageningen, Netherlands, International Livestock Research Institute (ILRI).

Burrell, J. & Oreglia, E. 2013. *The myth of market price information: mobile phones and epistemology in ICTD.* Working Paper. Berkeley, USA, University of California (available at https://markets.ischool.berkeley.edu/files/2013/03/MythOfMarketPrice_wp.pdf).

Byerlee, D. & Fischer, K. 2002. Accessing modern science: policy and institutional options for agricultural biotechnology in developing countries. *World Development,* 30(6): 931–958.

Cavatassi, R., Lipper, L. & Narloch, U. 2010. Modern variety adoption and risk management in drought prone areas: insights from the sorghum farmers of eastern Ethiopia. *Agricultural Economics,* 42(3): 279–292.

CIAT. 2012. *LINKing Smallholders: a guide on inclusive business models.* Website (available at http://dapa.ciat.cgiar.org/linking-smallholders-a-guide-on-inclusive-business-models/).

Classen, L., Humphries, S., Fitzsimons, J., Kaaria, S., Jiménez, J., Sierra, F. & Gallardo, O. 2008. Opening participatory spaces for the most marginal: learning from collective action in the Honduran hillsides. *World Development,* 36(11): 2402–2420.

Collier, P. 2008. The politics of hunger: how illusion and greed fan the food crisis. *Foreign Affairs,* 87(6): 67–79.

Critchley, W., Reij, C. & Willcocks, T. 1994. Indigenous soil and water conservation: a review of the state of knowledge and prospects for building on traditions. *Land Degradation and Development,* 5(4): 293–314.

Dasgupta, P. & Maler, K. 1995. Poverty, institutions and the environmental resource base. *In* J. Behrman & T. Srinivisan, *Handbook of development economics, Volume IIIB.* Amsterdam, North-Holland Publishing.

Davis, K. 2008. Extension in sub-Saharan Africa: overview and assessment of past and current models, and future prospects. *Journal of International Agricultural and Extension Education,* 15(3): 15-28.

Davis, K. & Place, N. 2003. Non-governmental organizations as an important actor in agricultural extension in semiarid east Africa. *Journal of International Agricultural and Extension Education,* 10(1): 31–36.

Davis, K., Ekboir, J. & Spielman, D. 2008. Strengthening agricultural education and training in sub-Saharan Africa from an innovation systems perspective: a case study of Mozambique. *The Journal of Agricultural Education and Extension,* 14(1): 35–51.

Davis, K., Swanson, B., Amudavi, D., Ayalew Mekonnen, D., Flohrs, A., Riese, J., Lamb, C. & Zerfu, E. 2010. *In-depth assessment of the public agricultural extension system of Ethiopia and recommendations for improvement.* IFPRI Discussion Paper 01041. Washington, DC, IFPRI.

De Soto, H. 2002. *The other path: the economic answer to terrorism*. New York, USA, Basic Books.

Deininger, K., Jin, S. & Nagarajan, H. 2009. Determinants and consequences of land sales market participation: panel evidence from India. *World Development,* 37(2): 410–421.

Deller, S. & Preissing, J. 2008. *The specialist in today's University of Wisconsin – Extension.* Agriculture and Applied Economics Staff Paper No. 521. Madison, USA, University of Wisconsin-Madison.

Doss, C.R. & Morris, M. 2001. How does gender affect the adoption of agricultural innovations? The case of improved maize technology in Ghana. *Agricultural Economics,* 25(1): 27–39.

Eastwood, R., Lipton, M. & Newell, A. 2010. Farm size. *In* P. Pingali & R. Evenson, eds. *Handbook of agricultural economics,* Vol. 4, Chapter 65, pp. 3323–3394. Amsterdam, North Holland.

Echeverría, R. & Beintema, N. 2009. *Mobilizing financial resources for agricultural research in developing countries: trends and mechanisms.* Rome, Global Forum for Agricultural Research.

Economic Research Service (United States Department of Agriculture). 2013. International agricultural productivity. Online dataset (available at http://www.ers.usda.gov/data-products/international-agricultural-productivity.aspx).

Eicher, C. 2006. *The evolution of agricultural education and training: global insights of relevance for Africa.* Washington, DC, World Bank.

Ekboir, J. 2003. Research and technology policies in innovation systems: zero tillage in Brazil. *Research Policy,* 32(4): 573–586.

Ekboir, J., Dutrénit, G., Martínez, V., Torres Vargas, A. & Vera-Cruz, A. 2009. *Successful organizational learning in the management of agricultural research and innovation: the Mexican produce foundations*. IFPRI Research Report No. 162. Washington, DC, IFPRI.

Evenson, R. 2001. Economic impacts of agricultural research and extension. *In* B. Gardner & G. Rausser, eds. *Handbook of agricultural economics,* Vol. 1A, Chapter 11, pp. 573–628,. Amsterdam, North Holland.

Evenson, R. & Gollin, D. 2003. Assessing the impact of the Green Revolution, 1960 to 2000. *Science,* 300(5620): 758–762.

Fafchamps, M. & Minten, B. 2012, November. Impact of SMS-based agricultural information on Indian farmers. *World Bank Economic Review,* 26(3): 383–414.

Fan, S. & Chan-Kang, C. 2005. Is small beautiful? Farm size, productivity, and poverty in Asian agriculture. *Agricultural Economics,* 32(Issue Supplement s1): 135–146.

Fan, S., Brzeska, J., Keyzer, M. & Halsema, A. 2013. *From subsistence to profit. Transforming smallholder farms*, Food Policy Report. Washington, DC, IFPRI.

FAO. 1995. *World agriculture: towards 2010.* Rome.

FAO. 2001. *Supplement to the report on the 1990 World Census of Agriculture*. FAO Statistical Development Series 9a. Rome.

FAO. 2005a. *A system of integrated agricultural censuses and surveys.* Volume 1. *World Programme for the Census of Agriculture 2010.* Rome.

FAO. 2005b. *Annotated bibliography on and stage-wise analysis of participatory research projects in agriculture and natural resource management.* Rome.

FAO. 2006. *Technology for agriculture. Labour saving technologies and practices decision support tool.* Website (available at http://teca.fao.org/).

FAO. 2007. *The State of Food and Agriculture 2007. Paying farmers for environmental services.* Rome.

FAO. 2008a. FAOSTAT. Online statistical database (retrieved 2008) (available at http://faostat.fao.org).

FAO. 2008b. *Market-oriented agricultural infrastructure: appraisal of public–private partnerships.* Agricultural Management, Marketing and Finance Occasional Paper No. 23. Rome.

FAO. 2009. *How to feed the world in 2050.* Rome.

FAO. 2010a. *"Climate-smart" agriculture: policies, practices and financing for food security, adaptation and mitigation.* Rome.

FAO. 2010b. *Corporate Strategy on Capacity Development.* FAO Programme Committee Document PC104/3. Rome.

FAO. 2011a. *The State of the World's Land and Water Resources for Food and Agriculture. Managing systems at risk.* Rome.

FAO. 2011b. *The State of Food and Agriculture 2010–11. Women in agriculture: closing the gender gap for development.* Rome.

FAO. 2011c. *Save and grow: a policymaker's guide to the sustainable intensification of smallholder crop production.* Rome.

FAO. 2012a. *Report of the FAO Expert Consultation on agricultural innovation systems and family farming.* Rome (available at http://www.fao.org/docrep/015/an761e/an761e00.pdf).

FAO. 2012b. *The State of Food and Agriculture 2012. Investing in agriculture for a better future.* Rome.

FAO. 2012c. *An FAO e-mail conference on agricultural innovation systems and family farming: the moderator's summary.* Rome (available at http://www.fao.org/docrep/016/ap097e/ap097e00.pdf).

FAO. 2012d. *Experiencias y enfoques de procesos participativos de innovación en agricultura: el caso de la Corporación PBA en Colombia*. Estudios sobre Innovación en la Agricultura Familiar. Rome.

FAO. 2013a. *2000 World Census of Agriculture: analysis and international comparison of the results (1996–2005).* FAO Statistical Development Series No. 13. Rome.

FAO. 2013b. *International year of family farming 2014. Master plan.* Rome (available at http://www.fao.org/fileadmin/user_upload/iyff/docs/Final_Master_Plan_IYFF_2014_30-05.pdf).

FAO. 2013c. *Agribusiness public–private partnerships: a country report of Thailand.* Rome.

FAO. 2013d. FAOSTAT. Online statistical database (retrieved November 2013) (available at http://faostat.fao.org).

FAO, 2013e. *Ensuring small-scale farmers can benefit from high food prices. The implications of smallholder heterogeneity in market participation.* Rome.

FAO. 2013f. *Tropical agriculture platform: assessment of current capacities and needs for capacity development in agricultural innovation systems in low income tropical countries.* Rome.

FAO. 2014a. Smallholders data portrait (available at http://www.fao.org/economic/esa/esa-activities/esa-smallholders/dataportrait/en/).

FAO. 2014b. FAOSTAT. Online statistical database (retrieved November 2014) (available at http://faostat.fao.org).

FAO. 2014c. *Public expenditure.* Monitoring and analysing food and agricultural policies (MAFAP) online database (retrieved July, 2014) (available at http://www.fao.org/mafap/database/public-expenditure/en/).

FAO & IFAD. 2012. *Good practices in building innovative rural institutions to increase food security.* Rome.

FAO & OECD. 2012. *Sustainable agricultural productivity growth and bridging the gap for small-family farms. Interagency report to the Mexican G20 presidency. Co-ordinated by FAO and OECD, with contributions by Bioversity, CGIAR Consortium, FAO, IFAD, IFPRI, IICA, OECD, UNC.* Rome and Paris.

FARA & ANAFE (Forum for Agricultural Research in Africa and African Network for Agriculture, Agroforestry & Natural Resources Education). 2005. *BASIC: Building Africa's scientific and institutional capacity in agriculture and natural resources education.* Proceedings of a meeting of African Networks and Associations that Build Agricultural Capacity at Universities, 23–25 November 2005. Nairobi.

Farrington, J. & Martin, A. 1988. *Farmer participation in agricultural research: a review of concepts and practices.* Agricultural Administration Unit Occasional Paper No. 9. London, Overseas Development Institute.

Faure, G. & Kleene, P. 2002. Management advice for family farms in West Africa: role of the producers' organizations in the delivery of sustainable Agricultural Extension Services. Montpellier, France, CIRAD.

Feder, G., Murgai, R. & Quizon, J. 2003. *Sending farmers back to school: the impact of farmer field schools in Indonesia.* World Bank Policy Research Working Paper No. 3022. Washington, DC, World Bank.

Fuglie, K. 2012. Productivity growth and technology capital in the global agricultural economy. *In* K. Fuglie, S. Wang & V. Ball, eds. *Productivity growth in agriculture: an international perspective.* Wallingford, UK, Centre for Agriculture and Biosciences International (CABI).

Fuglie, K., Heisey, P., King, J., Pray, C., Day-Rubenstein, K., Schimmelpfennig, D., Ling Wang, S. & Karmarkar-Deshmukh, R. 2011. *Research investments and market structure in the food processing, agricultural input and biofuel industries worldwide*. Economic Research Report ERR-130. Washington, DC, United States Department of Agriculture, Economic Research Service.

Galli, A., Wiedmann, T., Ercin, E., Knoblauch, D., Ewinge, B. & Giljum, S. 2012. Integrating ecological, carbon and water footprint into a "footprint family" of indicators: definition and role in tracking human pressure on the planet. *Ecological Indicators,* 16 (May 2012): 100–112.

Garner, E. & de la O Campos, A. 2014. *Identifying the "family farm": an informal discussion of the concepts and definitions.* ESA Working Paper No. 14-10. Rome, FAO.

GFRAS. 2014. *Regional services.* Global Forum on Rural Advisory Services (available at http://

www.g-fras.org/en/weblinks/155-root/37-regional-services-and-initiatives.html).

Government of Brazil. 2009. *Censo Agropecuário 2006.* Rio de Janeiro, Instituto Brasileiro de Geografia e Estatística (IBGE).

Government of Lao People's Democratic Republic. 2012. *Lao Census of Agriculture 2010/11. Highlights.* Summary census report. Vientiane, Ministry of Agriculture and Forestry.

Government of Malawi. 2010. *National Census of Agriculture and Livestock 2006/07. Main report.* Zomba, Malawi, National Statistical Office.

Government of Nicaragua. 2012. *IV Censo Nacional Agropecuario (IV CENAGRO, 2011).* Managua, Instituto Nacional de Información de Desarrollo.

Government of Paraguay. 2009. *Censo Agropecuário Nacional 2008.* San Lorenzo, Ministero de Agricultura y Ganadería.

Government of Uganda. 2011. *Uganda Census of Agriculture 2008/09.* Kampala, Uganda Bureau of Statistics.

Graeub, B., Chappell, J., Wittman, H., Ledermann, S., Batello, C. & Gemmill-Herren, B. (forthcoming). *The state of family farmers in the world: global contributions and local insights for food security.* Rome, FAO.

Grameen Foundation. 2013a. *Community knowledge worker.* Webpage (retrieved September 2013) (available at http://www.grameenfoundation.org/what-we-do/agriculture/community-knowledge-worker).

Grameen Foundation. 2013b. *By the numbers.* Webpage (retrieved September 2013 (available at http://www.grameenfoundation.org/our-impact/numbers).

Graziano da Silva, J., Del Grossi, M.E. & de França, C.G., eds. 2010. *The Fome Zero (Zero Hunger Program): the Brazilian experience.* Brasilia, FAO and the Ministry of Agrarian Development.

Hall, A. & Dijkman, J. 2009. Will a time of plenty for agricultural research help to feed the world? *LINK Look* editorial, *Link news bulletin,* Nov.-Dec. 2009. Hyderabad, India, United Nations University.

Hall, A., Sulaiman, V. & Clark, N. & Yoganand, B. 2003. From measuring impact to learning institutional lessons: an innovation systems perspective on improving the management of international agricultural research. *Agricultural Systems,* 78(2): 213–241.

Hartwich, F., Tola, J., Engler, A., González, C., Ghezan, G., Vázquez-Alvarado, J.M.P., Silva, J.A., de Jésus Espinoza, J. & Gottret, M.V. 2008. *Food security in practice: building public–private partnerships for agricultural innovation.* Washington, DC, IFPRI.

Haverkort, B., van der Kamp, J. & Waters-Bayer, A. 1991. *Joining farmers' experiments: experiences in participatory development.* London, IT Publications.

Hayami, Y. & Ruttan, V. 1971. *Agricultural development. An international perspective.* Baltimore, MD, The Johns Hopkins Press.

Hazell, P.B. & Hess, U. 2010. Drought insurance for agricultural development and food security in dryland areas. *Food Security,* 2: 395–405.

Hazell, P., Poulton, C., Wiggins, S. & Dorward, A. 2010. The future of small farms: trajectories and policy priorities. *World Development,* 38(10): 1349–1361.

Heemskerk, W., Nederlof, S. & Wennink, B. 2008. *Outsourcing agricultural advisory services: enhancing rural innovation in sub-Saharan Africa.* Amsterdam, Royal Tropical Institute (KIT).

Herdt, R.W. 2012 People, institutions, and technology: a personal view of the role of foundations in international agricultural research and development 1960–2010. *Food Policy,* 37(2): 179–190.

HLPE. 2013. Investing in smallholder agriculture for food security. HLPE Report 6. A report by the High Level Panel of Experts on Food Security and Nutrition, Committee on World Food Security. Rome, FAO.

Hounkonnou, D., Kossou, D., Kuyper, T. & Leeuwis, C., Nederlof, E.S., Röling, N., Sakyi-Dawson, O., Traoré, M. & van Huis, A. 2012. An innovation systems approach to institutional change: smallholder development in West Africa. *Agricultural Systems,* 108: 74–83.

Humphries, S., Gallardo, O., Jimenez, J. & Sierra, F. 2005. *Linking small farmers to the formal research sector: lessons from a participatory bean breeding program in Honduras,* Network Paper No. 142. London, Agricultural Research & Extension Network (AgREN), Overseas Development Institute.

Hurley, T., Pardey, P. & Rao, X. 2013. *Returns to food and agricultural R&D investments worldwide 1958–2011.* INSTEPP Brief. Saint Paul, USA, University of Minnesota.

IFPRI (International Food Policy Research Institute). 2012. *Global Food Policy Report 2012.* Washington, DC.

IFPRI. 2013a. SPEED Data visualization tool. Online database (retrieved November 2013) (available at http://www.ifpri.org/tools/speed).

IFPRI. 2013b. *The status of food security in the feed the future zone and other regions of Bangladesh: results from the 2011–2012 Bangladesh Integrated Household Survey.* Washington, DC, USAID.

IMF (International Monetary Fund). 2013. Government finance statistics. Online database (retrieved November 2013) (available at http://elibrary-data.imf.org/FindDataReports.aspx?d=33061&e=170809).

IPCC (International Plant Protection Convention). 2007. Summary for policymakers. *In* S. Solomon, D. Qin, M. Manning, Z. Chen, M. Marquis, K. Averyt, M. Tignor & H.L. Miller, eds. *Climate change 2007: the physical science basis. Contribution of working group I to the Fourth assessment report of the Intergovernmental Panel on Climate Change.* Cambridge, UK, and New York, USA, Cambridge University Press.

IPPC. 2014. *Climate change 2014: impacts, adaptation and vulnerability. IPCC WGII AR5 Summary for policymakers.* Cambridge, UK, Cambridge University Press.

Jia, X. & Huang, J. 2013. *Transforming agricultural production in China: from smallholders to pluralistic large farms.* Rome, Presentation made at FAO headquarters on 16 December 2013.

Jiggins, J. & de Zeeuw, H. 1992. Participatory technology development in practice: process and methods. *In* C. Reijntje, B. Haverkort & A. Waters-Bayer, eds. *Farming for the future.* Netherlands, Macmillan and the Centre for Learning on Sustainable Agriculture (ILEIA).

Juma, C. 1987. *Ecological complexity and agricultural innovation: the use of indigenous genetic resources in Bungoma, Kenya.* Paper presented at the meeting on Farmers and Agricultural Research: Complementary Methods, 27–31 July 1987. Brighton, UK, Institute of Development Studies (IDS), University of Sussex.

Kahan, D. 2007. *Farm management extension services: a review of global experience.* Agricultural Management, Marketing and Finance Occasional Paper No. 21. Rome, FAO.

Kahan, D. 2011. *Market-oriented advisory services in Asia. A review and lessons learned.* Bangkok, FAO.

Karfakis, P., Ponzini, G. & Rapsomanikis, G. 2014 (forthcoming). *On the costs of being small: case evidence from Kenyan family farms.* ESA Working Paper No. 14-11. Rome, FAO.

Kidd, A., Lamers, J., Ficarelli, P. & Hoffmann, V. 2000. Privatising agricultural extension: caveat emptor. *Journal of Rural Studies,* 16(1): 95–102.

Kilpatrick, S. 2005. *Education and training: impacts on farm management practice.* Gosford, Australia, Centre for Research and Learning in Regional Australia, University of Tasmania.

Kiptot, E. & Franzel, S. 2014. Voluntarism as an investment in human, social and financial capital: evidence from a farmer-to-farmer extension program in Kenya. *Agriculture and Human Values*, 31: 231–243.

Kiptot, E., Franzel, S. & Kirui, J. 2012. *Volunteer farmer trainers: improving smallholder farmers' access to information for a stronger dairy sector.* Policy Brief No. 13. Nairobi, World Agroforestry Centre.

Kjær, A. & Joughin, J. 2012. The reversal of agricultural reform in Uganda: ownership and values. *Policy and Society,* 31(4): 319–330.

Klerkx, L. & Gildemacher, P. 2012. The role of innovation brokers in agricultural innovation systems. *In* World Bank. *Agricultural innovation systems: an investment sourcebook,* Module 3, Thematic Note 4. Washington, DC.

Klerkx, L., Aarts, N. & Leeuwis, C. 2010. Adaptive management in agricultural innovation systems: the interactions between innovation networks and their environment. *Agricultural Systems,* 103(6): 390–400.

Klerkx, L., Hall, A. & Leeuwis, C. 2009. Strengthening agricultural innovation capacity: are innovation brokers the answer? *International Journal of Agricultural Resources, Governance and Ecology*, 8(5–6): 409–438.

Larson, D., Otsuka, K., Matsumoto, T. & Kilic, T. 2013. *Should African rural development strategies depend on smallholder farms? An exploration of the inverse productivity hypothesis.* Policy Research Paper No. 6190. Washington, DC, World Bank.

Leeuwis, C. & Van den Ban, A. 2004. *Communication for rural innovation: rethinking agricultural extension.* Oxford, UK, Blackwell Science.

Lipton, M. 2006. Can small farmers survive, prosper, or be the key channel to cut mass poverty? *Electronic Journal of Agricultural and Development Economics,* 3(1): 58–85.

Long, N. & Long, A. 1992. *Battlefields of knowledge: the interlocking of theory and practice in social research and development.* London, Routledge.

Lowder, S., Skoet, J. & Singh, S. 2014. *What do we really know about the number and distribution of farms, family farms and farmland worldwide?* Background paper for *The State of*

Food and Agriculture 2014. ESA Working Paper No. 14-02. Rome, FAO.

Masters, W., Andersson Djurfeldt, A., De Haan, C., Hazell, P., Jayne, T., Jirstrom, M. & Reardon, T. 2013. Urbanization and farm size in Asia and Africa: implications for food security and agricultural research. *Global Food Security,* 2(3): 156–165.

McCarthy, N., Lipper, L. & Branca, G. 2011. *Climate smart agriculture: smallholder adoption and implications for climate change adaptation and mitigation.* Mitigation of Climate Change in Agriculture (MICCA) Working Paper No. 4. Rome, FAO.

Meinzen-Dick, R., Johnson, N., Quisumbing, A.R., Njuki, J., Berhman, J.A., Rubin, D., Peterman, A. & Waithanji, E. 2014. The gender asset gap and its implications for agricultural and rural development. *In* A. Quisumbing, R. Meinzen-Dick, T. Raney, A. Croppenstedt, J. Behrman & A. Peterman, eds. *Gender in agriculture: closing the knowledge gap.* Rome, FAO, and Washington, DC, Springer Science/IFPRI.

Meinzen-Dick, R., Quisumbing, A., Behrman, J., Biermayr-Jenzano, P., Wilde, V., Noordeloos, M., Ragasa, C. & Beintema, N. 2011. *Engendering agricultural research, development and extension.* Washington, DC, IFPRI.

Millennium Ecosystem Assessment. 2005. *Ecosystems and human well-being: synthesis.* Washington, DC, Island Press.

Mogues, T., Morris, M., Freinkman, L., Adubi, A. & Ehui, S. 2008. *Agricultural public spending in Nigeria.* IFPRI Discussion Paper No. 00789. Washington, DC, IFPRI.

Mogues, T., Yu, B., Fan, S. & McBride, L. 2012. *The impacts of public investment in and for agriculture.* IFPRI Discussion Paper No. 01217. Washington, DC, IFPRI.

Nagel, J. 2010. *Acceso y uso de tics en pequeños agricultores.* Presentation at Taller CEGES, Chile, December.

Nederlof, S., Wongtschowski, M. & van der Lee, F. 2011. *Putting heads together: agricultural innovation platforms in practice.* Amsterdam, KIT.

Nelson, G., van der Mensbrugghe, D., Ahammad, H., Blanc, E., Calvin, K., Hasegawa, T., Havlik, P., Heyhoe, E., Kyle, P., Lotze-Campen, H., von Lampe, M., d'Croz, D.M., van Meijl, H., Müller, C., Reilly, J., Robertson, R., Sands, R.D., Schmitz, C., Tabeau, A., Takahashi, K., Valin, H. & Willenbockel, D. 2014. Agriculture and climate change in global scenarios: why don't the models agree. *Agricultural Economics,* 45(1): 85–101.

News China Magazine, 2013 (April). China promotes family farms. Online news article (retrieved on 13 May 2014) (available at http://www.newschinamag.com/magazine/china-promotes-family-farms).

Nie, F. & Fang, C. 2013. *Family farming in China: structural changes, government policies and market development for growth inclusive of smallholders.* Rome, Presentation made at FAO headquarters on 13 December 2013.

OECD & Eurostat. 2005. *Oslo manual: guidelines for collecting and interpreting innovation data,* third edition. Oslo, Organisation for Economic Co-operation and Development (OECD).

OECD. 2006. *The challenge of capacity development. Working towards good practice.* DAC Guidelines and Reference Series. Paris, OECD.

OECD. 2013. *Agricultural innovation systems: a framework for analysing the role of the government.* Paris, OECD.

OECD & FAO. 2012. *OECD–FAO Agricultural Outlook 2012–2021.* Paris and Rome.

OECD & FAO. 2014. *OECD–FAO Agricultural Outlook 2014–2023.* Paris and Rome.

Padgham, P. 2009. *Agricultural development under a changing climate: opportunities and challenges for adaptation,* Joint Discussion Paper, Issue 1. Washington, DC, World Bank.

Pal, S., Rahija, M. & Beintema, N. 2012. *India: recent development in agricultural research.* ASTI Country Note. Washington, DC, and New Delhi, IFPRI & Indian Council of Agricultural Research (ICAR).

Pandolfelli, L., Meinzen-Dick, R. & Dohrn, S. 2008. Introduction, gender and collective action: motivations, effectiveness and impact. *Journal of International Development,* 20(1): 1–11.

Pardey, P. & Beddow, J. 2013. *Agricultural innovation: the United States in a changing global reality.* Chicago, USA, The Chicago Council on Global Affairs.

Pardey, P. & Beintema, N. 2001. *Slow magic.* Food Policy Report No. 13. Washington, DC, IFPRI.

Pardey, P., Alston, J. & Ruttan, V. 2010. The economics of innovation and technical change in agriculture. *In* B. Hall & N. Rosenberg, eds. *Handbook of the economics of innovation,* Vol. 2, Chapter 22. New York, USA, Elsevier.

Pardey, P., Chan-Kang, C. & Dehmer, S. 2014. *Global food and agricultural R&D spending, 1960–2009.* InSTePP Report. St Paul, USA, University of Minnesota.

Phillips, P., Karwandy, J., Webb, G. & Ryan, C. 2013. *Innovation in agri-food clusters: theory and case* *studies.* Wallingford, UK, Centre for Agriculture and Biosciences International, CABI Publishing.

Place, F. & Meybeck, A. 2013. *Food security and sustainable resource use: what are the resource challenges to food security?* Background paper for the conferenceon Food Security Futures, Research Priorities for the 21st Century, Dublin, April 2013.

Posthumus, H., Martin, A. & Chancellor, T. 2012. *A systematic review on the impacts of capacity strengthening of agricultural research systems for development and the conditions of success.* London, Evidence for Policy and Practice Information and Co-ordinating Centre (EPPI–Centre), Social Science Research Unit, Institute of Education, University of London.

Poulton, C. & Kanyinga, K. 2013. *The politics of revitalising agriculture in Kenya.* Future Agricultures Working Paper 059. Brighton, UK, Future Agricultures Consortium (FAC).

Power, A. 2010. Ecosystem services and agriculture: tradeoffs and synergies. *Philosophical Transactions of the Royal Society B, Biological Sciences,* 365(1554): 2959–2971.

Pray, C. & Nagarajan, L. 2012. *Innovation and research by private agribusiness in India,* IFPRI Discussion Paper No. 1181. Washington, DC, IFPRI.

Preissing, J. 2012. INCAGRO: Developing a market for agricultural innovation services in Peru. *In* World Bank. *Agricultural innovation systems: an investment sourcebook.* Washington, DC.

Pretty, J. 2008. Agricultural sustainability: concepts, principles and evidence. *Philosophical Transactions of the Royal Society B, Biological Sciences,* 363(1491): 447–465.

Pretty, J., Noble, A., Bossio, D., Dixon, J., Hine, R., de Vries, F. & Morison, L. 2006. Resource-conserving agriculture increases yields in developing countries. *Environmental Science & Technology,* 40(4): 1114–1119.

Pretty, J., Toulmin, C. & William, S. 2011. Sustainable intensification in African agriculture. *International Journal of Agricultural Sustainability,* 9(1): 3–4.

Proctor, F. & Lucchesi, V. 2012. *Small-scale farming and youth in an era of rapid rural change.* London and The Hague, International Institute for Environment and Development (IIED) and Humanist Institute for Development Cooperation (Hivos).

PROLINNOVA. 2012. *Farmer access to innovation resources findings and lessons learnt on facilitating local innovation support fund.* Leusden, Netherlands, Promoting Local Innovation in Ecologically Oriented Agriculture and Natural Resource Management (PROLINNOVA) International Secretariat.

Raabe, K. 2008. *Reforming the agricultural extension system in India: what do we know about what works where and why?* IFPRI Discussion Paper No. 775. Washington, DC, IFPRI.

Ragasa, C., Sengupta, D., Osorio, M., OurabahHaddad, N. & Mathieson, K. 2014. *Gender-specific approaches and rural institutions for improving access to and adoption of technological innovation.* Rome, FAO.

Rajalahti, R., Janssen, W. & Pehu, E. 2008. *Agricultural innovation systems: from diagnostics toward operational practices.* Agriculture and Rural Development Discussion Paper No. 38, Washington, DC, World Bank.

Raney, T. 2006. Economic impact of transgenic crops in developing countries. *Current Opinion in Biotechnology,* 17(2): 174–178.

Rao, X., Hurley, T. & Pardey, P. 2012. *Recalibrating the reported rates of return to food and agricultural R&D.* Staff Paper P12–8. St Paul, Minnesota, USA, University of Minnesota, Department of Applied Economics.

Rapsomanikis, G. 2014. *The economic lives of smallholder farmers,* Rome, FAO.

Rausser, G., Simon, L. & Ameden, H. 2000. Public–private alliances in biotechnology: can they narrow the knowledge gaps between rich and poor? *Food Policy,* 25(4): 499–513.

Reardon, T. & Timmer, C. 2012. The economics of the food system revolution. *Annual Review of Resource Economics,* 4: 225–264.

Reijntjes, C., Haverkort, B. & Waters-Bayer, A. 1992. *Farming for the future.* Netherlands, Macmillan and Centre for Learning on Sustainable Agriculture (ILEIA).

Reimers, M. & Klasen, S. 2013. Revisiting the role of education for agricultural productivity. *American Journal of Agricultural Economics,* 95(1): 131–152.

Ricker-Gilbert, J., Norton, G., Alwang, J., Miah, M. & Feder, G. 2008. Cost effectiveness of alternative pest management extension methods: an example from Bangladesh. *Review of Agricultural Economics,* 30(2): 252–269.

Rivera, W. 2011. Public sector agricultural extension system reform and challenges ahead. *Journal of Agricultural Education and Extension,* 17(2): 165–180.

Rivera, W. & Zijp, W., eds. 2002. *Contracting for agricultural extension: international case studies and emerging practices.* New York, USA, CABI Publishing.

Rodrigues, M. & Rodríguez, A. 2013. *Information and communication technologies for agricultural development in Latin America: trends, barriers and policies.* Santiago, Comisión Económica para América Latina (CEPAL).

Röling, N. & Engel, P. 1989. IKS and knowledge management: utilizing indigenous knowledge in institutional knowledge systems. *In* D.M. Warren, L. Jan Slikkerveer & S. Oguntunji Titilola, eds. *Indigenous knowledge systems: implications for agriculture and international development.* Studies in Technology and Social Change No. 11. Ames, USA, Technology and Social Change Program, Iowa State University.

Roseboom, J. 2012. Creating an enabling environment for agricultural innovation. *In* World Bank. *Agricultural innovation systems: an investment sourcebook.* Washington, DC.

Rwamigisa, B., Birner, R., Mangheni, M. & Arseni Semana, A. 2013. *How to promote institutional reforms in the agricultural sector? A case study of Uganda's National Agricultural Advisory Services (NAADS).* Paper presented at the International Conference on the Political Economy of Agricultural Policy in Africa. Pretoria,20–18 March 2013, organized by the Futures Agriculture Consortium and the Institute for Poverty, Land and Agrarian Studies (PLAAS).

Quisumbing, A. & Pandolfelli, L. 2010. Promising approaches to address the needs of poor female farmers: resources, constraints, and interventions. *World Development, 38* (4): 581–592.

Schultz, T. 1964. *Transforming traditional agriculture.* Chicago, USA, University of Chicago Press.

Schumpeter, J. 1939. *Business cycles: a theoretical, historical and statistical analysis of the capitalist process,* New York, McGraw-Hill.

Scoones, I. & Thompson, J., eds. 1994. *Beyond Farmer First: rural people's knowledge, agricultural research and extension practice,* London, IT Publications.

Shah, N. & Jansen, F. 2011. *Digital alternatives with a cause.* Bangalore, India, Centre for internet and society, and The Hague, Netherlands, Hivos Knowledge Programme.

Singh, S.P., Puna Ji Gite, L. & Agarwal, N. 2006. Improved farm tools and equipment for women workers for increased productivity and reduced drudgery. *Gender, Technology and Development,* 10 (2): 229–244.

Sitko, N. 2010. Study presented at the Agro-enterprise learning alliance for southern and eastern Africa. Michigan State University, Michigan, USA.

Spielman, D., Hartwich, F. & von Grebmer, K. 2007. *Public–private partnerships in international agricultural research.* Research Brief No. 9, Washington, DC, IFPRI.

Spielman, D. & Birner, R. 2008. *How innovative is your agriculture? Using innovation indicators and benchmarks to strengthen national agricultural innovation systems.* Agriculture and rural development discussion paper No. 41. Washington, DC, World Bank.

Stads, G.-J. 2011. *Africa's agricultural R&D funding rollercoaster. An analysis of the elements of funding volatility.* ASTI/IFPRI-FARA Conference Working Paper 2. Prepared for the Agricultural Science and Technology Indicators (ASTI), IFPRI, and Forum for Agricultural Research in Africa (FARA) Conference on Agricultural R&D, Investing in Africa's Future, Accra, Ghana, 5–7 December 2011.

Starkey, P.S. 2002. *Improving rural mobility: options for developing motorized and non motorized transport in rural areas.* World Bank Technical Paper No. 525, Washington, DC, World Bank.

Sulaiman, R. & Hall, A. 2002. *Beyond technology dissemination: can Indian agricultural extension re-invent itself?* Policy Brief No. 16. New Delhi, National Centre for Agricultural Economics and Policy Research.

Swanson, B. & Rajalahti, R. 2010. *Strengthening agricultural extension and advisory systems: procedures for assessing, transforming, and evaluating extension systems.* Agriculture and Rural Development Discussion Paper No. 4. Washington, DC, World Bank.

Swanson, B., Farner, B. & Bahal, R. 1988. *Report of the global consultation on agricultural extension: the current status of agricultural extension worldwide.* Rome, FAO.

Tewes-Gradl, C., Peters, A.Vohla, K. & Lütjens-Schilling, L. 2013. *Inclusive business policies: how governments can engage companies in meeting development goal .* Berlin, Enterprise Solutions for Development (Endeva).

Thapa, S. 2008. *Gender differentials in agricultural productivity: evidence from Nepalese household data.* Munich Personal RePEc Archive (MPRA) Paper No. 13722 (available at http://mpra.ub.unimuenchen.de/13722/).

World Bank, Development Prospects Group. 2013. World Bank commodity price data (The Pink Sheet) (retrieved November 2013) (available at worldbank.org).

Thiele, G., Devaux, A., Reinoso, I., Pico, H., Montesdeoca, F., Pumisacho, M. & Manrique, K. 2009. November. Multi-stakeholder platforms for innovation and coordination in market chains. In *15th Triennial International Symposium of the International Society for Tropical Root Crops (ISTRC).*

Thomas, C., Cameron, A., Bakkenes, M., Beaumont, L., Collingham, Y.C., Green, R.E., Erasmus, B., Ferreira de Siqueira, M., Grainger, A., Hannah, L., Hughes, L., Huntley, B., van Jaarsveld, A., Midgley, G., Miles, L., Ortega-Huerta, M., Townsend Peterson, A., Phillips, O. & Williams, S. 2004. Extinction risk from climate change. *Nature,* 427(6970): 145–148.

Thompson, J., Porras, I.T., Tumwine, J.K., Mujwahuzi, M.R., Katui-Katua, M., Johnstone, N. & Wood, L. 2001. *Drawers of water II: 30 years of change in domestic water use and environmental health in East Africa.* Summary. London, UK, International Institute for Environment and Development.

Thornton, P. & Lipper, L. 2013. *How does climate change alter agricultural strategies to support food security?* Dublin, Ireland, Background paper for the conference on Food security futures: research priorities for the 21st Century, 11 - 12 April 2013.

Ton, G., de Grip, K., Klerkx, L., Rau, M-L., Douma, M., Friis-Hansen, E., Triomphe, B., Waters-Bayer, A. & Wongtschowski, M. 2013. *Effectiveness of innovation grants to smallholder agricultural producers: an explorative systematic review.* London, Evidence for Policy and Practice Information and Co-ordinating Centre (EPPI-Centre), Social Science Research Unit, Institute of Education, University of London.

Triomphe, B., Floquet, A., Kamau, G., Letty, B., Vodouhe, S.D., Ng'ang'a, T., Stevens, J., van den Berg, J., Selemna, N., Bridier, B., Crane, T., Almekinders, C., Waters-Bayer, A. & Hocdé, H. 2013. What does an inventory of recent innovation experiences tell us about agricultural innovation in Africa? *The Journal of Agricultural Education and Extension,* :(3)19 324–311.

Tschirley, D., Minde, I. & Boughton, D. 2009. *Contract farming in sub-Saharan Africa: lessons from cotton on what works and under what conditions.* Issues Brief No. 7, Pretoria, Regional Strategic Analysis and Knowledge Support System (RESAKSS).

Udry, C., Hoddinott, J., Alderman, H. & Haddad, L. 1995. Gender differentials in farm productivity: implications for household efficiency and agricultural policy. *Food Policy,* 20(5): 407–423.

Umali, D. & Schwartz, L. 1994. *Public and private agricultural extension beyond traditional frontiers.* Washington, DC, World Bank.

UNDP (United Nations Development Programme). 2008. *Creating value for all: strategies for doing business with the poor.* New York, USA.

UNDP. 2010. *The MDGs. Everyone's business: how inclusive business models contribute to development and who supports them.* New York, USA.

United Nations. 2011. *World Economic and Social Survey 2011: the great green technological transformation.* New York, USA.

United Nations. 2013. *World population prospects: the 2012 revision.* New York, USA.

Van Campenhout, B. 2012, June 15. *Mobile apps to deliver extension to remote areas: preliminary results from Mnt Elgon area.* Grameen Foundation (available at http://www.grameenfoundation.org/resource/mobile-applications-deliver-extension-remote-areas).

Vernooy, R., Shrestha, P., Ceccarelli, S., Labrada, H.R., Song, Y. & Humphries, S. 2009. Towards new roles, responsibilities and rules: the case of participatory plant breeding. *In* S. Ceccarelli, E. Guimarães & E. Weltzien, eds. *Plant breeding and farmer participation,* pp. 613–671. Rome, FAO.

Viala, E. 2008. Water for food, water for life. A comprehensive assessment of water management in agriculture. *Irrigation and Drainage Systems,* 22(1): 127–129.

Vollan, B. 2012. Pitfalls of externally initiated collective action: a case study from South Africa. *World Development,* 40(4): 758–770.

von Lampe, M., Willenbockel, D., Ahammad, H., Blanc, E., Cai, Y., Calvin, K., Fujimori, S., Hasegawa, T., Havlik, P., Heyhoe, E., Kyle, P., Lotze-Campen, H., d'Croz, D.M., Nelson, G.C., Sands, R.D., Schmitz, C., Tabeau, A., Valin, H., van der Mensbrugghe, D. & van Meijl, H. 2014. Why do global long-term scenarios for agriculture differ? An overview of the AgMIP global economic model intercomparison. *Agricultural Economics,* 45(1): 3–20.

Wennink, B. & Heemskerk, W. 2006. *Farmers' organizations and agricultural innovation: case* *studies from Benin, Rwanda and Tanzania.* Amsterdam, Royal Tropical Institute (KIT).

Wettasinha, C., Wongtschowski, M. & Waters-Bayer, A. 2008. *Recognising and enhancing local innovation.* PROLINNOVA Working Paper No. 13. Leusden, Netherlands, PROLINNOVA Secretariat, ETC EcoCulture, Silang, International Institute of Rural Reconstruction.

WFP & FAO. 2007. *Getting started! Running a junior farmer field and life school.* Rome, FAO & WFP.

World Bank. 2006. *Enhancing agricultural innovation: how to go beyond the strengthening of research systems.* Washington, DC.

World Bank. 2007a. *Cultivating knowledge and skills to grow African agriculture.* A synthesis of an Institutional, regional and international review. Washington, DC.

World Bank. 2007b. *Philippines: Agriculture Public Expenditure Review.* Working Paper No. 40493, Washington, DC.

World Bank. 2007c. *World Development Report 2008: Agriculture for development.* Washington, DC.

World Bank. 2008. *Agricultural innovation systems: from diagnostics toward operational practices.* Washington, DC.

World Bank. 2009. *Agribusiness and innovation systems in Africa.* Washington, DC.

World Bank. 2010a. *Indonesia: Agriculture Public Expenditure Review.* Washington, DC.

World Bank. 2010b. *Innovation policy: a guide for developing countries.* Washington, DC.

World Bank. 2012a. *World Development Indicators 2012.* Washington, DC.

World Bank. 2012b. *Agricultural innovation systems: an investment sourcebook.* Washington, DC.

World Bank. 2013. *World Development Indicators* database (available at http://data.worldbank.org/data-catalog/world-development-indicators/wdi-2013).

Wright, B. & Pardey, P. 2006. Changing intellectual property regimes: implications for developing country agriculture. *International Journal for Technology and Globalization,* 2(1/2): 93–114.

Yorke, L., 2009. *Grameen Foundation launches mobile services tailored to the poor with Google and MTN Uganda.* Grameen Foundation (retrieved 18 September 2013) (available at http://www.kiwanja.net/media/docs/Grameen-Foundation-AppLab-Release.pdf).

Special chapters of *The State of Food and Agriculture*

Each issue of this report since 1957 has included one or more special studies on problems of longer-term interest. Special chapters in earlier issues have covered the following subjects:

1957	Factors influencing the trend of food consumption
	Postwar changes in some institutional factors affecting agriculture
1958	Food and agricultural developments in Africa south of the Sahara
	The growth of forest industries and their impact on the world's forests
1959	Agricultural incomes and levels of living in countries at different stages of economic development
	Some general problems of agricultural development in less-developed countries in the light of postwar experience
1960	Programming for agricultural development
1961	Land reform and institutional change
	Agricultural extension, education and research in Africa, Asia and Latin America
1962	The role of forest industries in the attack on economic underdevelopment
	The livestock industry in less-developed countries
1963	Basic factors affecting the growth of productivity in agriculture
	Fertilizer use: spearhead of agricultural development
1964	Protein nutrition: needs and prospects
	Synthetics and their effects on agricultural trade
1966	Agriculture and industrialization
	Rice in the world food economy
1967	Incentives and disincentives for farmers in developing countries
	The management of fishery resources
1968	Raising agricultural productivity in developing countries through technological improvement
	Improved storage and its contribution to world food supplies
1969	Agricultural marketing improvement programmes: some lessons from recent experience
	Modernizing institutions to promote forestry development
1970	Agriculture at the threshold of the Second Development Decade
1971	Water pollution and its effects on living aquatic resources and fisheries
1972	Education and training for development
	Accelerating agricultural research in the developing countries
1973	Agricultural employment in developing countries
1974	Population, food supply and agricultural development
1975	The Second United Nations Development Decade: mid-term review and appraisal
1976	Energy and agriculture
1977	The state of natural resources and the human environment for food and agriculture
1978	Problems and strategies in developing regions
1979	Forestry and rural development
1980	Marine fisheries in the new era of national jurisdiction
1981	Rural poverty in developing countries and means of poverty alleviation
1982	Livestock production: a world perspective
1983	Women in developing agriculture
1984	Urbanization, agriculture and food systems
1985	Energy use in agricultural production

	Environmental trends in food and agriculture
	Agricultural marketing and development
1986	Financing agricultural development
1987–88	Changing priorities for agricultural science and technology in developing countries
1989	Sustainable development and natural resource management
1990	Structural adjustment and agriculture
1991	Agricultural policies and issues: lessons from the 1980s and prospects for the 1990s
1992	Marine fisheries and the law of the sea: a decade of change
1993	Water policies and agriculture
1994	Forest development and policy dilemmas
1995	Agricultural trade: entering a new era?
1996	Food security: some macroeconomic dimensions
1997	The agroprocessing industry and economic development
1998	Rural non-farm income in developing countries
2000	World food and agriculture: lessons from the past 50 years
2001	Economic impacts of transboundary plant pests and animal diseases
2002	Agriculture and global public goods ten years after the Earth Summit
2003–04	Agricultural biotechnology: meeting the needs of the poor?
2005	Agriculture trade and poverty: can trade work for the poor?
2006	Food aid for food security?
2007	Paying farmers for environmental services
2008	Biofuels: prospects, risks and opportunities
2009	Livestock in the balance
2010–11	Women in agriculture: closing the gender gap for development
2012	Investing in agriculture for a better future
2013	Food systems for better nutrition